THE DANCING WU LI MASTERS

gary zukav

THE
DANCING
WU LI
MASTERS

An Overview
of the New Physics

PERENNIAL 📕 CLASSICS

HarperCollins books may be purchased for educational, business, or sales
promotional use. For information please write: Special Markets Department,
HarperCollins Publishers, Inc., 10 East 53rd Street, New York, NY 10022.

First Quill edition published 1979.
First Perennial Classics edition published 2001.
Perennial Classics are published by Perennial, an imprint of HarperCollins
Publishers

Designed by Stanley S. Drate/Folio Graphics Co. Inc.

Library of Congress Cataloging-in-Publication Data
Zukav, Gary
    The dancing wu li masters : an overview of the new physics / Gary Zukav.
      p. cm.
    Originally published: New York : Morrow, 1979.
    Includes bibliographical references and index.
    ISBN 0-681-18965-7
    1. Quantum theory—History.   2. Relativity (Physics)—History I. Title.

QC173.98. Z84   2001
530.12—dc21   2001021386

---

05 06 07 RRD 1 2 3 4 5 6 7 8 9 10

This book is dedicated to you, who
are drawn to read it.

Acknowledgments

My gratitude to the following people cannot be adequately expressed. I discovered, in the course of writing this book, that physicists, from graduate students to Nobel Laureates, are a gracious group of people; accessible, helpful, and engaging. This discovery shattered my long-held stereotype of the cold, "objective" scientific personality. For this, above all, I am grateful to the people listed here.

Jack Sarfatti, Ph.D., brought me together with many physicists. Al Chung-liang Huang, The T'ai Chi Master, provided the perfect metaphor of *Wu Li,* inspiration, and the beautiful calligraphy. David Finkelstein, Ph.D., Director of the School of Physics, Georgia Institute of Technology, was my first tutor.

In addition to Finkelstein and Sarfatti, Brian Josephson, Professor of Physics, Cambridge University, and Max Jammer, Professor of Physics, Bar-Ilan University, Ramat-Gan, Israel, read and commented upon the entire manuscript. I am especially indebted to these men (but I do not wish to imply that any one of them, or any other of the individualistic and creative thinkers who helped me with this book, would approve of it, page for page, as it is written, nor that the responsibility for any errors or misinterpretations belongs to anyone but me).

x 🕉 Acknowledgments

I am also indebted to Henry Stapp, Ph.D., Lawrence Berkeley Laboratory, for reading and commenting upon portions of the manuscript, and to Elizabeth Rauscher, Ph.D., founder and sponsor of the Fundamental Physics Group, Lawrence Berkeley Laboratory, for encouraging non-physicists to partake of weekly conferences which normally would attract only physicists. In addition to Stapp and Sarfatti, this group included John Clauser, Ph.D.; Philippe Eberhard, Ph.D.; George Weissman, Ph.D.; Fred Wolf, Ph.D.; and Fritjof Capra, Ph.D.; among others.

I am grateful to Carson Jefferies, Professor of Physics, University of California at Berkeley, for his support and for commenting upon portions of the manuscript; to David Bohm, Professor of Physics, Birkbeck College, University of London, for reading portions of the manuscript; to Saul-Paul Sirag for his frequent assistance; to the physicists of the Particle Data Group, Lawrence Berkeley Laboratory, for their assistance in compiling the particle table at the back of the book; to Eleanor Criswell, Professor of Psychology, Sonoma State University (California), for her valuable support; to Gin McCollum, Professor of Mathematics, Kansas State University, for her assistance and for her patient tutelage; and to Nick Herbert, Ph.D., Director of the C-Life Institute, for providing me with his excellent papers on Bell's theorem and for permission to use his paper title, "More Than Both," as a chapter title.

All of the illustrations in this book were done by Thomas Linden Robinson.

Harvey White, Professor Emeritus, Department of Physics, University of California at Berkeley, and former Director, Lawrence Hall of Science, personally provided photographs of his famous simulations of probability distribution patterns. The electron diffraction photograph was provided by Ronald Gronsky, Ph.D., Lawrence Berkeley Laboratory. I learned much about spectroscopy from Sumner Davis, Professor of Physics, University of California at Berkeley. I am deeply grateful to these men who, like all of the physicists that I encountered in writing this book, gave graciously of their time and knowledge to a stranger who needed help.

I also am indebted to Maria Guarnaschelli, my editor, for her sensitivity and erudition.

Without the generosity of Michael Murphy and the Board of Directors of the Esalen Institute, which sponsored the 1976 conference on physics and consciousness, none of this probably would have happened.

Contents

Synoptic Table of Contents   *xv*

Cast of Characters   *xix*

Foreword by David Finkelstein   *xxiii*

Introduction to the Perennial Classics Edition   *xxvii*

Introduction   *xxxi*

WU LI?

Big Week at Big Sur   *3*

Einstein Doesn't Like It   *19*

PATTERNS OF ORGANIC ENERGY

Living?   *49*

What Happens   *74*

MY WAY

The Role of "I"   *101*

NONSENSE
Beginner's Mind   *131*
Special Nonsense   *150*
General Nonsense   *179*

I CLUTCH MY IDEAS
The Particle Zoo   *213*
The Dance   *236*

ENLIGHTENMENT
More Than Both   *283*
The End of Science   *312*

Notes   *349*
Bibliography   *357*
Index   *362*
Stable Particle Table   *378–79*

Synoptic Table of Contents

WU LI? (Introduction)

Big Week at Big Sur

Physics (3), Esalen (4), Chinese and English (5–6), Wu Li Masters (7), scientists and technicians (10), the sodium spectrum (11–12), Bohr's model of the atom (14).

Einstein Doesn't Like It

The new physics and the old physics (20), Newton's physics (22), the Great Machine (24), do we create reality? (30), the myth of objectivity (32), subatomic "particles" (34), statistics (35), the kinetic theory of gases (36), probability (37), the Copenhagen Interpretation of Quantum Mechanics (40), pragmatism (41), split-brain analysis (42), summary of the new physics and the old physics (44).

PATTERNS OF ORGANIC ENERGY
(Quantum Mechanics)

Living?

Organic and inorganic (49), Max Planck (52), "discontinuous" (53), black-body radiation (54), Planck's constant (56), Albert Einstein

(57), Einstein's theory of the photoelectric effect (58), waves, wavelengths, frequencies, and amplitudes (60), diffraction (63), Young's double-slit experiment (66), the wave-particle duality (70), probability waves (72).

What Happens

The procedure of quantum mechanics (74), the region of preparation (75), the region of measurement (75), the observed system (76), the observing system (76), the Schrödinger wave equation (77), observables (77), particles as "correlations" (78), wave functions (80), probability functions (81), quantum jumps (83), the Theory of Measurement (87), the metaphysics of quantum mechanics (88), the Many Worlds Interpretation of Quantum Mechanics (92), Schrödinger's cat (94), Doubting Thomas (97).

MY WAY (Quantum Mechanics)

The Role of "I"

The "in here—out there" illusion (102), complementarity (103), Compton scattering (103), Louis de Broglie (106), matter waves (107), Erwin Schrödinger (110), standing waves (110), the Pauli exclusion principle (114), the Schrödinger wave equation (again) (114), Max Born (117), probability waves (again) (117), the quantum model of the atom (119), Werner Heisenberg (121), the S Matrix (122), the Heisenberg uncertainty principle (123), the tables are turned (127).

NONSENSE (Relativity)

Beginner's Mind

Nonsense (131), the beginner's mind (132), the special theory of relativity (134), the Galilean relativity principle (138), inertial co-ordinate systems (138), Galilean transformations (140), the constancy of the speed of light (142), the ether (145), the Michelson-Morley experiment (145), FitzGerald contractions (148), Lorentz transformations (148).

Special Nonsense

The special theory of relativity (150), "proper" and "relative" length and time (155), Terrell's rotation explanation of relativistic contraction (159), relativistic mass increase (162), simultaneity (162), the space-time continuum (167), the space-time interval (171), Hermann Minkowski (173), mass-energy (173), conservation laws (176).

General Nonsense

Gravity and acceleration (181), inside and outside the elevators (181), gravitational mass and inertial mass (186), the geography of the space-time continuum (188), Euclidean geometry (191), the revolving circles (193), non-Euclidean geometry (196), Einstein's ultimate vision (200), Mercury's perihelion (201), starlight deflection (203), gravitational redshift (204), Black Holes (205), the illusion of "force" (208), the illusion of "nonsense" (209).

I CLUTCH MY IDEAS (Particle Physics)

The Particle Zoo

The barriers to change (213), the hall of mirrors (215), the new world view (215), particle physics (216), bubble chambers (218), the dance of creation and annihilation (219), what made the tracks? (221), quantum field theory (222), the need to pretend (224), particle masses (226), massless particles (228), charge (229), spin (230), angular momentum (231), quantum numbers (234), anti-particles (235).

The Dance

Space-time diagrams (237), Feynman diagrams (239), the dance of creation and annihilation (again) (240), anti-particles (again) (242), the illusion of time (245), entropy (246), virtual photons (247), the electromagnetic force (251), Hideki Yukawa (252), the strong force (252), virtual mesons (254), self-interactions (254), gravity (260), the weak force (260), virtual photons (again) (261), vacuum diagrams (266), conservation laws (269), symmetries (271), quarks (272), the S Matrix (again) (272).

ENLIGHTENMENT
(Quantum Logic & Bell's Theorem)
More Than Both

Physics and enlightenment (283), Bell's theorem and quantum logic (285), John von Neumann (286), the description of a wave function (286), "Projections as Propositions" (288), David Finkelstein (290), symbols and experience (290), logos and mythos (290), the distributive law (292), polarization of light (293), the third polarizer paradox (297), superpositions (299), quantum logic (301), "proof" (301), transition tables (303), lattices (305), von Neumann's disproof of the distributive law (302), quantum topology (311).

The End of Science

Enlightenment and unity (312), J. S. Bell (313), quantum connectedness (313), the Einstein-Podolsky-Rosen thought experiment (314), superluminal communication (319), the principle of local causes (320), Bell's theorem (322), the Freedman-Clauser experiment (323), the Aspect experiment (327), contrafactual definiteness (332), superdeterminism (333), the Many Worlds Theory (again) (333), summary (335), the philosophy of quantum mechanics (338), David Bohm (339), unbroken wholeness (339), implicate order (340), the "new" thought instrument (341), eastern psychologies (342), the metaphor of physics (343), Kali (345), the Path without Form (347), the circle dance (348).

Cast of Characters

Thomas Young
  1803 (double-slit experiment)
Albert Michelson, Edward Morley
  1887 (Michelson-Morley experiment)
George Francis FitzGerald
  1892 (FitzGerald contractions)
Hendrik Antoon Lorentz
  1893 (Lorentz transformations)
Electron
  1897 (discovered)
Max Planck
  1900 (quantum hypothesis)
Albert Einstein
  1905 (photon theory)
  1905 (special theory of relativity)
Hermann Minkowski
  1908 (space-time)
Nucleus
  1911 (discovered)
Niels Bohr
  1913 (specific-orbits model of the atom)

Albert Einstein
   1915 (general theory of relativity)
Louis de Broglie
   1924 (matter waves)
Niels Bohr, H. A. Kramers, John Slater
   1924 (first concept of probability waves)
Wolfgang Pauli
   1925 (exclusion principle)
Werner Heisenberg
   1925 (matrix mechanics)
Erwin Schrödinger
   1926 (Schrödinger wave equation)
   1926 (equates matrix mechanics with wave mechanics)
   1926 (visits Bohr in Copenhagen to attack the idea of quantum
        jumps—and gets the flu)
Max Born
   1926 (probability interpretation of wave function)
Niels Bohr
   1927 (complementarity)
Clinton Davisson, Lester Germer
   1927 (Davisson-Germer experiment)
Werner Heisenberg
   1927 (uncertainty principle)
Copenhagen Interpretation of Quantum Mechanics
   1927
Paul Dirac
   1928 (anti-matter)
Neutron
   1932 (discovered)
Positron
   1932 (discovered)
John Von Neumann
   1932 (quantum logic)
Albert Einstein, Boris Podolsky, Nathan Rosen
   1935 (EPR paper)
Hideki Yukawa
   1935 (predicts meson)

Meson
    1947 (discovered)
Richard Feynman
    1949 (Feynman diagrams)
Sixteen New Particles
    1947–1954 (discovered)
Many Worlds Interpretation of Quantum Mechanics
    1957
David Finkelstein
    1958 (one-way membrane hypothesis)
Quasars
    1962 (discovered)
Quarks
    1964 (hypothesized)
J. S. Bell
    1964 (Bell's theorem)
David Bohm
    1970 (implicate order)
Henry Stapp
    1971 (nonlocal connections re: Bell's theorem)
Stuart Freedman, John Clauser
    1972 (Freedman-Clauser experiment)
Twelve New Particles
    1974–1977 (discovered)
Alain Aspect
    1982 (Aspect experiment)

Foreword

When Gary Zukav announced his plans for this book, creating the out-
line with Al Huang and me watching at a dinner table at Esalen, 1976,
I did not realize the magnitude of the job he took on with such joy.
Watching the book grow has been instructive and rewarding, because
Zukav has insisted on going through the whole evolution of the quantum relativistic physics of today, treating it as it is, an unfolding story.
As a result this book is not only readable, but it also puts the reader
in touch with all the various ways that physicists have worked out for
talking about what is so hard to talk about. In short, Gary Zukav has
written a very good book for laymen.

Zukav's attitude to physics is rather close to mine, so I must be
a layman too, and it is more stimulating to talk physics with him than
with most professionals. He knows that physics is—among other
things—an attempt to harmonize with a much greater entity than ourselves, requiring us to seek, formulate and eradicate first one and then
another of our most cherished prejudices and oldest habits of
thought, in a never-ending quest for the unattainable.

Zukav has graciously offered me this place to add my own
emphases to his narrative. Since it has been three years since we met,
I must sift my memory for a while.

Migrating whales come to mind first. I remember us standing on the Esalen cliffs and watching them cavort as they headed south. Next comes to mind beautiful Monarch butterflies, dotting the fields from the first day, and covering one magic tree as thick as leaves in a grand finale. Between the whales and the butterflies it was difficult for us to feel self-important and very easy for us to play.

The very difficulty of communicating with the physicists at Esalen helped me to realize how differently most physicists think about quantum mechanics than I do. Not that my way is new, it is one of two ways already pointed out in John Von Neumann's 1932 book, *The Mathematical Foundation of Quantum Mechanics:*

1. Quantum mechanics deals with propositions defined by processes of preparation and observation involving subject and object and obeying a new logic; not with objective properties of the object alone.
2. Quantum mechanics deals with objective properties of the object alone, obeying the old logic, but they jump in a random way when an observation is made.

Most working physicists seem to see one of these ways (the second) and not the other. Perhaps personality can determine the direction of science. I think there are "thing" minds and "people" minds. Good parents, psychologists and writers have to be "people" people, while mechanics, engineers and physicists tend to be "thing" people. Physics has become too scary for such physicists because it is already so thingless. New evolutions, as profound as those of Einstein and Heisenberg, are waiting for a new generation of more daring and integrated thinkers.

While most physicists take for granted the quantum tools of their daily work, there is a vanguard already testing roads to the next physics, and a rearguard still conscientiously holding the road back to the old. Bell's theorem is mainly important to the latter, and its prominence in the book does not mean it uncovers problems in present-day quantum physics. Rather Bell's theorem drives toward a view that most physicists already assume: that quantum mechanics is something new and different.

Here it helps to distinguish between a *complete* theory, predicting everything, what Newtonians look for (it does not seem that Newton was a strict Newtonian, since he wanted God to reset the world clock now and then) and a *maximal* theory, predicting as much as possible, what quantum physicists look for. In spite of their controversy, Einstein and Bohr both agreed, in their different ways, that quantum mechanics is incomplete, and even that it is not yet maximal. What they really debated was whether or not an incomplete theory can be maximal. Throughout their famous controversy Einstein argued, "Alas, our theory is too poor for experience," and Bohr replied, "No, no! Experience is too rich for our theory"; just as some existential philosophers despair at the indeterminacy of life and the existence of choices, and others feel *élan vital*.

One of the features of quantum mechanics that leads to such controversy is its concern with the nonexistent, the potential. There is some of this in all language, or words could only be used once, but quantum mechanics is more involved with probabilities than classical mechanics. Some people feel this discredits quantum theory, makes it less than maximal theory. So it is important to mention in defense of quantum theory that in spite of indeterminacy, quantum mechanics can be entirely expressed in yes-or-no terms about individual experiments, just like classical mechanics, and that probabilities can be derived as a law of large numbers and need not be postulated. I prefer to state the difference between classical and quantum theories not as presented in textbooks, but thus: Once sufficient data is given, classical mechanics gives yes-or-no answers for all further questions while quantum mechanics simply leaves *unanswered* some questions in the theory, to be answered by experience. I wish here also to note the regrettable tendency, in myself also, to feel that quantum mechanics must thereby deny physical existence to those answers that are found in experience only, not in the theory, such as the momentum of a localized electron. So involved are we in our symbol systems.

After a week of talking, the conference was still working on the elements of quantum logic, and never did get far into the new quantum time concepts we wanted to try out, but it made it easier to move on to the next set of problems, which occupy me today. Quantum mechanics is characterized by its unanswered questions. Some logicians, Martin Davis for one, have suggested these may be related to the undecidable propositions dominating logic since Gödel. I used to know better. Nowadays I think they may be right, the common element being reflexivity and the impossibility for finite systems of total self-knowledge. The proper study of mankind is endless, it seems. I hope these ideas work out and Gary Zukav writes a book about them. He does it well.

DAVID FINKELSTEIN
New York
July 1978

Introduction to the Perennial Classics Edition

When I wrote The Dancing Wu Li Masters: An Overview of the New *Physics,* I had never written a book and I had never studied physics. In fact, I didn't like science and I had no mathematical aptitudes. Yet while I was writing *The Dancing Wu Li Masters,* I knew it would be published and that it would be very well received. I also knew that it would sell very well for many years after its publication. I did not need to have faith in these things. I knew them. I could see them. It was clear to me that all I needed to do to make them happen was to continue writing. In other words, to do was my part. I knew that without my part, none of what I saw would happen, and that with my part, it was already accomplished.

I was the key. Everything depended only upon my doing what I was already doing—writing about physics, studying physics, discussing physics, and writing about it again each day. That was no problem for me because I loved doing those things. I woke thinking about the ideas in *The Dancing Wu Li Masters* and I went to sleep thinking about them the same way that some people wake in the morning and go to sleep at night thinking about a Beloved. Every decision about what word or words to use, what ideas to include, and how to present a discussion was made with the reader in mind. "The reader," no matter whom I pictured in that role, was always someone who was intelligent—perhaps more intelligent than I. He or she was keenly interested in all that I had to share, but had no background in science or mathematics.

I do not know how I knew these things while I was writing *The Dancing Wu Li Masters*—that it would be published, that it would be well received, and that it would sell around the world for a very long time—but they were realities to me. I also knew that I would not remain interested in physics indefinitely, even though it was my passion at the moment. Because of that, I decided to leave behind me the best gift that I could for those who would come later. A spirit of giving pervades this book, and that is one reason why, I believe, *The Dancing Wu Li Masters* has been so appreciated by more than a million readers, in many languages, around the world.

Another reason is that *The Dancing Wu Li Masters* contains within it the seed of the thought that consciousness lies at the heart of all that we can experience, all that we can conceive, and all that we are. It also points toward the possibility that intentions create the reality that we experience. These thoughts were pondered by many of the founders of quantum mechanics as they discussed complementarity, the Uncertainty Principle and other aspects of the mathematical formalism that became known as the quantum theory. They are still pondered by philosophically inclined physicists today.

When I began writing *The Dancing Wu Li Masters*, I could only suspect or deduce things about the role of consciousness and intention in the creation of experience, but I soon discovered that some of the founders of quantum mechanics had suspected and deduced the same things decades before me. This was exciting enough, yet as I became more engaged in the writing, I began to encounter phenomena that I never could have imagined. For example, I realized that the book that I was writing was more intelligent than I was. It was also funnier than I was, and it had a grander comprehension than I did. For example, I prepared an outline for each chapter before I began to write. The outline contained the central idea of the chapter, what I wanted to include in the chapter, how I intended to order the content of the chapter, and how I intended to present the ideas in the chapter.

In every instance, I would write not for long before I had to choose between following the outline that I had prepared and going with the energy that developed as I wrote. I always went with the energy. If I had planned a certain discussion, but another one occurred to me that felt more exciting, I used the second discussion. I substituted examples, words, and ideas as they occurred to me for those that I had put into my outline. And I was amazed at the experience of a chapter coming to an end—at how clear it was for me, and how often the ending of a chapter surprised and delighted me.

As I wrote more chapters, I noticed something else, too. The chapters fit together perfectly—even though I had not planned them that way. I might have been able to take credit for this if I had followed each of the outlines that I had prepared. But I never followed my outlines. I always followed the flow of energy and excitement that I felt as I wrote.

Who orchestrated this? Who planned for a chapter—one that I completed before I began research on a later chapter—to fit into a later chapter as though I had written them together? Where did the humor in the book come from—the humor that took me away from the torment of my daily judgements about myself and others? Where did the gratitude originate that replaced my worries about paying the rent, which obsessed me when I was not writing?

Eventually, the reality of these miracles became a part of my awareness. So did the contrast between the painful life that I lived when I was not working on *The Dancing Wu Li Masters* and the fulfillment that I felt when I was. At last, I decided to live my entire life the way that *The Dancing Wu Li Masters* was being written—spontaneously, intelligently, and joyfully. Eventually I learned how to do this, and how to explain how to do it.

At the time I did not have the vocabulary to articulate what I was experiencing, or the ability to understand it. It was not until later that I was able to understand that writing *The Dancing Wu Li Masters* was my first experience of authentic power—of meaning, fulfillment, and purpose. It was also my first experience of nonphysical assistance.

Now there is a vocabulary to express what I experienced. It is the vocabulary of authentic power—the alignment of the personality with the soul—and of multisensory perception: the expansion of human perception beyond the limitations of the five senses. I could not see at the time that my experiences were part of an evolutionary transformation that was, and still is, reshaping human experience. This transformation and the vocabulary to express it are the content of *The Seat of the Soul,* which I wrote ten years after *The Dancing Wu Li Masters* was published.

This same evolutionary transformation continues to reshape human experience, one human at a time. As multisensory perception—the ability to access information that the five senses cannot provide—emerges in millions of humans, they naturally become interested in the relationship between consciousness and physical reality. *The Dancing Wu Li Masters* is a nourishing book for these individuals, and will continue to be, because it explores that relationship with enthusiasm and integrity.

The last reason, I believe, that *The Dancing Wu Li Masters* is still so popular is that it is fun to read. It is an enjoyable way to learn about quantum mechanics, especially for people who have no mathematical or scientific background. It brings the early history of the quantum theory to life in a refreshing and exciting way, and that history will not change. The development of the quantum theory is one of the most well-documented intellectual adventures in the history of humanity. For those who come to this history as a newcomer, as I did, without preparation or prior interest, as I did, but with an open and excited mind, as mine was, *The Dancing Wu Li Masters* is the perfect book.

This Perennial Classics edition celebrates the contributions that *The Dancing Wu Li Masters* has made for more than two decades to those who delight in the rigors of intellectual exploration into the most meaningful questions that philosophy and science can pose. It is a product of that delight.

I hope that you enjoy it.

With Love,
Gary Zukav

Introduction

My first exposure to quantum physics occurred a few years ago when a friend invited me to an afternoon conference at the Lawrence Berkeley Laboratory in Berkeley, California. At that time, I had no connections with the scientific community, so I went to see what physicists were like. To my great surprise, I discovered that (1) I understood everything that they said, and (2) their discussion sounded very much like a theological discussion. I scarcely could believe what I had discovered. Physics was not the sterile, boring discipline that I had assumed it to be. It was a rich, profound venture which had become inseparable from philosophy. Incredibly, no one but physicists seemed to be aware of this remarkable development. As my interest in and knowledge of physics grew, I resolved to share this discovery with others. This book is a gift of my discovery. It is one of a series.

Generally speaking, people can be grouped into two categories of intellectual preference. The first group prefers explorations which require a precision of logical processes. These are the people who become interested in the natural sciences and mathematics. They do not become scientists because of their education, they choose a scientific education because it gratifies their scientific mental set. The second group prefers explorations which involve the intellect in a less logically rigorous manner. These are the people who become interested in the liberal arts. They do not have a liberal arts mentality because of their education, they choose a liberal arts education because it gratifies their liberal arts mental set.

Since both groups are intelligent, it is not difficult for members of one group to understand what members of the other group are studying. However, I have discovered a notable *communication* problem between the two groups. Many times my physicist friends have attempted to explain a concept to me and, in their exasperation, have tried one explanation after another, each one of which sounded (to me) abstract, difficult to grasp, and generally abstruse. When I could comprehend, at last, what they were trying to communicate, inevitably I was surprised to discover that the idea itself was actually quite simple. Conversely, I often have tried to explain a concept in terms which seemed (to me) laudably lucid, but which, to my exasperation, seemed hopelessly vague, ambiguous, and lacking in precision to my physicist friends. I hope that this book will be a useful *translation* which will help those people who do not have a scientific mental set (like me) to understand the extraordinary process which is occurring in theoretical physics. Like any translation, it is not as good as the original work and, of course, it is subject to the shortcomings of the translator. For better or worse, my first qualification as a translator is that, like you, I am not a physicist.

To compensate for my lack of education in physics (and for my liberal arts mentality) I asked, and received, the assistance of an extraordinary group of physicists. (They are listed in the acknowledgments). Four of them in particular, read the entire manuscript. As each chapter was completed, I sent a copy of it to each physicist and asked him to correct any conceptual or factual errors which he found. (Several other physicists read selected chapters.)

My original intention was to use these comments to correct the text. However, I soon discovered that my physicist friends had given more attention to the manuscript than I had dared to hope. Not only were their comments thoughtful and penetrating, but, taken together, they formed a significant volume of information by themselves. The more I studied them, the more strongly I felt that I should share these comments with you. Therefore, in addition to correcting the manuscript with them, I also included in the footnotes those comments which do not duplicate the corrected text. In particular, I footnoted those comments which would have slowed the flow of the text or made it technical, and those comments which disagreed with the text and also disagreed with the comments of the other physicists. By publishing dissenting opinions in the footnotes, I have been able to include numerous ideas which would have lengthened and complicated the book if they had been presented in the text. From the beginning of *The Dancing Wu Li Masters* to the end, no term is used which is not explained immediately before or after its first use. This rule is not followed in the footnotes. This gives the footnotes an unmitigated freedom of expression. However, it also means that the footnotes contain terms that are not explained before, during, or after their use. The text respects your status as newcomer to a vast and exciting realm. The footnotes do not.

However, if you read the footnotes as you read the book, you will have the rare opportunity to see what four of the finest physicists in the world have to say about it as they, in effect, read it along with you. Their footnotes punctuate, illustrate, annotate, and jab at everything in the text. Better than it can be described, these footnotes reveal the aggressive precision with which men of science seek to remove the flaws from the work of a fellow scientist, even if he is an untrained colleague, like me, and the work is nontechnical, like this book.

The "new physics," as it is used in this book, means quantum mechanics, which began with Max Planck's theory of quanta in 1900, and relativity, which began with Albert Einstein's special theory of relativity in 1905. The old physics is the physics of Isaac Newton, which he discovered about three hundred years ago. "Classical physics" means any physics that attempts to explain reality in such a manner that for every element of physical reality there is a corresponding element in the theory. Therefore, "classical physics" includes the physics of Isaac Newton and relativity, both of which are structured in this one-to-one manner. It does not, however, include quantum mechanics, which, as we shall see, is one of the things that makes quantum mechanics unique.

Be gentle with yourself as you read. This book contains many rich and multifaceted stories, all of which are heady (pun?) stuff. You cannot learn them all at once any more than you can learn the stories told in *War and Peace, Crime and Punishment,* and *Les Misérables* all at once. I suggest that you read this book for your pleasure, and not to learn what is in it. There is a complete index at the back of the book and a good table of contents in the front. Between the two of them, you can return to any subject that catches your interest. Moreover, by enjoying yourself, you probably will remember more than if you had set about to learn it all.

One last note; this is not a book about physics and eastern philosophies. Although the poetic framework of *Wu Li* is conducive to such comparisons, this book is about quantum physics and relativity. In the future I hope to write another book specifically about physics and Buddhism. In view of the eastern flavor of *Wu Li,* however, I have included in this book those similarities between eastern philosophies and physics that seemed to me so obvious and significant that I felt that I would be doing you a disservice if I did not mention them in passing.

Happy reading.

GARY ZUKAV
San Francisco
July 1978

Most of the fundamental ideas of science are essentially simple, and may, as a rule, be expressed in a language comprehensible to everyone.

—Albert Einstein[1]

Even for the physicist the description in plain language will be a criterion of the degree of understanding that has been reached.

—Werner Heisenberg[2]

If you cannot—in the long run—tell everyone what you have been doing, your doing has been worthless.

—Erwin Schrödinger[3]

Part One

WU LI?

1

🌀

Big Week at Big Sur

When I tell my friends that I study physics, they move their heads from side to side, they shake their hands at the wrist, and they whistle, "Whew! That's difficult." This universal reaction to the word "physics" is a wall that stands between what physicists do and what most people think they do. There is usually a big difference between the two.

Physicists themselves are partly to blame for this sad situation. Their shop talk sounds like advanced Greek, unless you are Greek or a physicist. When they are not talking to other physicists, physicists speak English. Ask them what they do, however, and they sound like the natives of Corfu again.

On the other hand, part of the blame is ours. Generally speaking, we have given up trying to understand what physicists (and biologists, etc.) really do. In this we do ourselves a disservice. These people are engaged in extremely interesting adventures that are not that difficult to understand. True, *how* they do what they do sometimes entails a technical explanation which, if you are not an expert, can produce an involuntary deep sleep. *What* physicists do, however, is actually quite simple. They wonder what the universe is really made of, how it works, what we are doing in it, and where it is going, if it is going anyplace at all. In short, they do the same things that we do on starry nights when we look up at the vastness of the universe and feel overwhelmed by it and a part of it at the same time. That is what physicists really do, and the clever rascals get paid for doing it.

Unfortunately, when most people think of "physics," they think of chalkboards covered with undecipherable symbols of an unknown mathematics. The fact is that physics is not mathematics. Physics, in essence, is simple wonder at the way things are and a divine (some call it compulsive) interest in how that is so. Mathematics is the *tool* of physics. Stripped of mathematics, physics becomes pure enchantment.

I had spoken often to a friend, who was a physicist, about the possibility of writing a book, unencumbered with technicalities and mathematics, to explain the exciting insights that motivate current physics. So when he invited me to a conference on physics that he and Michael Murphy were arranging at the Esalen Institute, I accepted with a purpose.

The Esalen Institute (it is named for an Indian tribe) is in northern California. The northern California coast is an awesome combination of power and beauty, but nowhere so much as along the Pacific Coast Highway between the towns of Big Sur and San Luis Obispo. The Esalen facilities are located about a half hour south of Big Sur between the highway and the coastal mountains on the one side and rugged cliffs overlooking the Pacific Ocean on the other. A dancing stream divides the northern third of the grounds from the remainder. On that side is a big house (called the Big House) where guests stay and groups meet, along with a small home where Dick Price (cofounder of Esalen with Murphy) stays with his family. On the other side of the stream is a lodge where meals are served and meetings are held, accommodations for guests and staff, and hot sulfur baths.

Dinner at Esalen is a multidimensional experience. The elements are candlelight, organic food, and a contagious naturalness that is the essence of the Esalen experience. I joined two men who already were eating. One was David Finkelstein, a physicist from Yeshiva University (in New York) who was attending the conference on physics. The other was Al Chung-liang Huang, a T'ai Chi Master who was leading a workshop at Esalen. I could not have chosen better companions.

The conversation soon turned to physics.

"When I studied physics in Taiwan," said Huang, "we called it Wu Li (pronounced 'Woo Lee'). It means 'Patterns of Organic Energy.'"

Everyone at the table was taken at once by this image. Mental lights flashed on, one by one, as the idea penetrated. "Wu Li" was more than poetic. It was the best definition of physics that the conference would produce. It caught that certain something, that living quality that we were seeking to express in a book, that thing without which physics becomes sterile.

"Let's write a book about Wu Li!" I heard myself exclaim. Immediately, ideas and energy began to flow, and in one stroke all of the prior planning that I had done went out the window. From that pooling of energy came the image of the Dancing Wu Li Masters. My remaining days at Esalen and those that followed were devoted to finding out what Wu Li Masters are, and why they dance. All of us sensed with excitement and certitude that we had discovered the channel through which the very things that we wanted to say about physics would flow.

The Chinese language does not use an alphabet like western languages. Each word in Chinese is depicted by a character, which is a line drawing. (Sometimes two or more characters are combined to form different meanings.) This is why it is difficult to translate Chinese into English. Good translations require a translator who is both a poet and a linguist.

For example, "Wu" can mean either "matter" or "energy." "Li" is a richly poetic word. It means "universal order" or "universal law." It also means "organic patterns." The grain in a panel of wood is Li. The organic pattern on the surface of a leaf is also Li, and so is the texture of a rose petal. In short, Wu Li, the Chinese word for physics, means "patterns of organic energy" ("matter/energy"

[Wu] + "universal order/organic patterns" [Li]. This is remarkable since it reflects a world view which the founders of western science (Galileo and Newton) simply did not comprehend, but toward which virtually every physical theory of import in the twentieth century is pointing! The question is not, "Do they know something that we don't?" The question is, "How do they know it?"

English words can be pronounced almost any way without changing their meanings. I was five years a college graduate before I learned to pronounce "consummate" as an adjective (con-SUM-mate). (It means "carried to the utmost extent or degree; perfect"). I live in anguish when I think of the times that I have spoken of *con*summate linguists, *con*summate scholars, etc. Someone always seemed to be holding back a smile, almost. I learned later that these were the people who read dictionaries. Nonetheless, my bad pronunciation never prevented me from being understood. That is because inflections do not change the denotation of an English word. "No" spoken with a rising inflection ("No?"), with a downward inflection ("No!"), and with no inflection ("No . . .") all mean (according to the dictionary) "a denial, a refusal, negative."

This is not so in Chinese. Most Chinese syllables can be pronounced several different ways. Each different pronunciation is a different word which is written differently and which has a meaning of its own. Therefore, the same syllable, pronounced with different inflections, which unaccustomed western listeners scarcely can distinguish, constitutes distinctly separate words, each with its own ideogram and meaning, to a Chinese listener. In English, which is an atonal language, these different ideograms are all written and pronounced the same way.

For example, there are over eighty different "Wu"s in Chinese, all of which are spelled and pronounced the same way in English. Al Huang has taken five of these "Wu"s, each of which, when combined with "Li," produces a different "Wu Li," each with the same English spelling, and each pronounced (in English) "Woo Lee."

The first Wu Li means "Patterns of Organic Energy." This is the Chinese way of saying "physics." (Wu means "matter" or "energy").

The second Wu Li means "My Way." (Wu means "mine" or "self.")

The third Wu Li means "Nonsense."(Wu means "void" or "nonbeing.")

The fourth Wu Li means "I Clutch My Ideas." (Wu means "to make a fist" or "clutch with a closed hand.")

The fifth Wu Li means "Enlightenment." (Wu means "enlightenment" or "my heart/my mind.")

If we were to stand behind a master weaver as he begins to work his loom, we would see, at first, not cloth, but a multitude of brightly colored threads from which he picks and chooses with his expert eye, and feeds into the moving shuttle. As we continue to watch, the threads blend one into the other, a fabric appears, and on the fabric, behold! A pattern emerges.

In a similar manner, Al Huang has created a beautiful tapestry from his own epistemological loom:

PHYSICS = WU LI
Wu Li = Patterns of Organic Energy
Wu Li = My Way
Wu Li = Nonsense
Wu Li = I Clutch My Ideas
Wu Li = Enlightenment

Each of the physicists at the conference, to a person, reported a resonance with this rich metaphor. Here, at last, was the vehicle through which we could present the seminal elements of advanced physics. By the end of the week, everyone at Esalen was talking about Wu Li.

At the same time that this was happening, I was trying to find out what a "Master" is. The dictionary was no help. All of its definitions involved an element of control. This did not fit easily into our image of the Dancing Wu Li Masters. Since Al Huang is a T'ai Chi Master, I asked him.

"That is the word that other people use to describe me," he said. To Al Huang, Al Huang was just Al Huang.

Later in the week, I asked him the same question again, hoping to get a more tangible answer.

"A Master is someone who started before you did," was what I got that time.

My western education left me unable to accept a nondefinition for my definition of a "Master," so I began to read Huang's book, *Embrace Tiger, Return to Mountain*. There, in the foreword by Alan Watts, in a paragraph describing Al Huang, I found what I sought. Said Alan Watts of Al Huang:

> He begins from the center and not from the fringe. He imparts an understanding of the basic principles of the art before going on to the meticulous details, and he refuses to break down the t'ai chi movements into a one-two-three drill so as to make the student into a robot. The traditional way . . . is to teach by rote, and to give the impression that long periods of boredom are the most essential part of training. In that way a student may go on for years and years without ever getting the feel of what he is doing.[1]

Here was just the definition of a Master that I sought. A Master teaches essence. When the essence is perceived, he teaches what is necessary to expand the perception. The Wu Li Master does not speak of gravity until the student stands in wonder at the flower petal falling to the ground. He does not speak of laws until the student, of his own, says, "How strange! I drop two stones simultaneously, one heavy and one light, and *both* of them reach the earth at the same moment!" He does not speak of mathematics until the student says, "There must be a way to express this more simply."

In this way, the Wu Li Master dances with his student. The Wu Li Master does not teach, but the student learns. The Wu Li Master always begins at the center, at the heart of the matter. This is the approach that we take in this book. It is written for intelligent people who want to know about advanced physics but who are ignorant of its terminology and, perhaps, of its mathematics. *The Dancing Wu Li Masters* is a book of essence; the essence of quantum mechanics, quantum logic, special relativity, general relativity, and some new ideas that indicate the direction that physics seems to be moving. Of course, who can know where the future goes? The only surety is that what we think today will be a part of the past tomorrow. Therefore, this book deals not with knowledge, which is always past tense anyway, but with imagination, which is physics come alive, which is Wu Li.

One of the greatest physicists of all, Albert Einstein, was perhaps a Wu Li Master. In 1938 he wrote:

> Physical concepts are free creations of the human mind, and are not, however it may seem, uniquely determined by the external world. In our endeavor to understand reality we are somewhat like a man trying to understand the mechanism of a closed watch. He sees the face and the moving hands, even hears its ticking, but he has no way of opening the case. If he is ingenious he may form some picture of a mechanism which could be responsible for all the things he observes, but he may never be quite sure his picture is the only one which could explain his observations. He will never be able to compare his picture with the real mechanism and he cannot even imagine the possibility of the meaning of such a comparison.[2]

Most people believe that physicists are explaining the world. Some physicists even believe that, but the Wu Li Masters know that they are only dancing with it.

I asked Huang how he structures his classes.

"Every lesson is the first lesson," he told me. "Every time we dance, we do it for the first time."

"But surely you cannot be starting new each lesson," I said. "Lesson number two must be built on what you taught in lesson number one, and lesson three likewise must be built on lessons one and two, and so on."

"When I say that every lesson is the first lesson," he replied, "it does not mean that we forget what we already know. It means that what we are doing is always new, because we are always doing *it* for the first time."

This is another characteristic of a Master. Whatever he does, he does with the enthusiasm of doing it for the first time. This is the source of his unlimited energy. Every lesson that he teaches (or learns) is a first lesson. Every dance that he dances, he dances for the first time. It is always new, personal, and alive.

Isidor I. Rabi, Nobel Prize winner in Physics and the former Chairman of the Physics Department at Columbia University, wrote:

> We don't teach our students enough of the intellectual content of experiments—their novelty and their capacity for opening new fields. . . . My own view is that you take these things personally. You do an experiment because your own philosophy makes you want to know the result. It's too hard, and life is too short, to spend your time doing something because someone else has said it's important. You must feel the thing yourself . . .[3]

Unfortunately, most physicists are not like Rabi. The majority of them, in fact, *do* spend their lives doing what other people have told them is important. That was the point Rabi was making.

This brings us to a common misunderstanding. When most people say "scientist," they mean "technician." A technician is a highly trained person whose job is to apply known techniques and principles. He deals with the known. A scientist is a person who seeks to know the true nature of physical reality. He deals with the unknown.

In short, scientists discover and technicians apply. However, it is no longer evident whether scientists really discover new things or whether they *create* them. Many people believe that "discovery" is actually an act of creation. If this is so, then the distinction between scientists, poets, painters, and writers is not clear. In fact, it is possible that scientists, poets, painters, and writers are all members of the same family of people whose gift it is by nature to take those things which we call commonplace and to *re-present* them to us in such ways that our self-imposed limitations are expanded. Those people in whom this gift is especially pronounced, we call geniuses.

The fact is that most "scientists" are technicians. They are not interested in the essentially new. Their field of vision is relatively narrow; their energies are directed toward applying what is already known. Because their noses often are buried in the bark of a particular tree, it is difficult to speak meaningfully to them of forests. The case of the mysterious hydrogen spectrum illustrates the difference between scientists and technicians.

When a white light, such as sunlight, enters a glass prism, one of the most beautiful of phenomena occurs. Out the other side of the prism comes not white light, but every color in the rainbow from dark red to light violet, with orange, yellow, green, and blue in between. This is because white light is made of all these different colors. It is a combination, whereas red light contains only red light, green light contains only green light, etc. Isaac Newton wrote his famous *Optiks* about this phenomenon three hundred years ago. This display of colors is called a white-light spectrum. The spectral analysis of white light shows a complete spectrum because white light contains all of the colors that our eyes can see (and some that they cannot see, like infrared and ultraviolet).

However, not every spectral analysis produces a complete spectrum. If we take one of the chemical elements, for example, like sodium, cause it to emit light, and shine that light through a glass prism, we get only *part* of a complete spectrum.

If an object is visible in a dark room, it is emitting light. If it appears red, for example, it is emitting red light. Light is emitted by "excited" objects. Exciting a piece of sodium does not mean offering it tickets to the Super Bowl. Exciting a piece of sodium means adding some energy to it. One way of doing this is to heat it. When we shine the light emitted by excited (incandescent) sodium through a prism, or spectroscope, we do not obtain the full array of colors characteristic of white light, *but only parts of it.* In the case of sodium, we obtain two thin yellow lines.

We also can produce a negative image of the sodium spectrum by shining white light through sodium vapor to see what parts of the white light the sodium vapor absorbs. White light passing through sodium vapor and then through a spectroscope produces the whole rainbow of colors *minus* the two yellow lines emitted by incandescent sodium.

Either way, the sodium spectrum always produces the same distinct pattern. It may be composed of black lines on an otherwise complete spectrum of colors, or it may be composed of colored lines without the rest of the spectrum, but it always remains the same.* This pattern is the fingerprint of the element sodium. Each element emits (or absorbs) only specific colors. Likewise, each element produces a specific spectroscopic pattern which never varies.

Hydrogen is the simplest element. It seems to have only two components; a proton, which has a positive charge, and an electron, which has a negative charge. We must say "it seems to have" because there is not one person alive who has ever seen a hydrogen atom. If hydrogen atoms exist, millions of them can exist on a pinhead, so small are they calculated to be. "Hydrogen atoms" is a speculation about what is inside of the watch. We can say only that the existence of such entities nicely explains certain observations that would be very difficult to explain otherwise, barring explanations such as "the devil did it," which still may prove to be correct. (It is this kind of explanation that drove Galileo, Newton, and Descartes to create what is now modern science.)

At one time physicists thought that atoms were constructed in the following way: At the center of an atom is a nucleus, just as the sun is at the center of our solar system. In the nucleus is located almost all of the mass of the atom in the form of positively charged particles (protons) and particles about the same size as protons but without a

---

*In practice, some of the lines representing transitions between higher energy states do not appear in absorption spectra.

charge (neutrons). (Only hydrogen has no neutrons in its nucleus.) Orbiting about the nucleus, as the planets orbit the sun, are electrons, which have almost no mass compared with the nucleus. Each electron has one negative charge. The number of electrons is always the same as the number of protons, so that the positive and negative charges cancel each other and the atom, as a whole, has no charge.

The problem with comparing this model of the atom with our solar system is that the distances between an atomic nucleus and its electrons are enormously greater than we picture the distances between the sun and its planets. The space occupied by an atom is so huge, compared with the mass of its particles (almost all of which is in the nucleus), that the electrons orbiting the nucleus are "like a few flies in a cathedral" according to Ernest Rutherford, who created this model of the atom in 1911.

This is the familiar picture of the atom that most of us learned in school, usually under duress. Unfortunately, this picture is obsolete, so you can forget the whole thing. We will discuss later how physicists currently think of an atom. The point here is that the planetary model of the atom formed the background against which a most puzzling problem was solved.

The spectrum of hydrogen, the simplest of the atoms, contains over one hundred lines! The patterns of the other elements are even more complicated. When we shine the light from excited hydrogen gas through a spectroscope, we get over one hundred different lines of color in a distinct pattern.* The question is, "How can such a simple thing like a hydrogen atom, which has only two components, a proton and an electron, account for such a complex spectrum?"

One way of thinking about light is to ascribe wave-like properties

---

*Accurately speaking, different experimental equipment is required to pho-tograph each series of the hydrogen spectrum. Therefore, most single photographs of the hydrogen spectrum show only about 10 lines. Theoretically, there are an infinite number of lines in each atomic spectrum. In fact, theoretically, there are an infinite number of lines in each series of each spectrum because the lines in the higher frequency range of each series become so closely spaced that, in effect, they form a continuum.

to it, and then to say that different colors have different frequencies, just as different sounds, which also are waves, have different frequencies. Arnold Sommerfield, a German physicist who also was an accomplished pianist, observed, tongue-in-cheek, that hydrogen atoms, which emit over one hundred different frequencies, must be more complicated than grand pianos, which emit only eighty-eight different frequencies!

It was a Danish physicist named Niels Bohr who came up with an explanation (in 1913) that made so much sense that it won him a Nobel Prize. Like most ideas in physics, it is essentially simple. Bohr did not start with what was theoretically "known" abut the structure of atoms. He started with what he really *knew* about atoms, that is, he started with raw spectroscopic data. Bohr speculated that electrons revolve around the nucleus of an atom not at just any distance, but in orbits, or shells, which are specific distances from the nucleus. Each of these shells (theoretically there are an infinite number of them), contains up to a certain number of electrons, but no more.

If the atom has more electrons than the first shell can accommodate, the electrons begin to fill up the second shell. If the atom has more electrons than the first and second shells combined can hold, the third shell begins to fill, and so on, like this:

Shell number	1	2	3	4	5 . . .
Numbers of electrons	2	8	18	32	50 . . .

His calculations were based on the hydrogen atom, which has only one electron. According to Bohr's theory, the electron in the hydrogen atom stays as close to the nucleus as it can get. In other words, it usually is in the first shell. This is the lowest energy state of a hydrogen atom. (Physicists call the lowest energy state of any atom its "ground state.") If we excite an atom of hydrogen we cause its electron to jump to one of the outer shells. How far it jumps depends upon how much energy we give it. If we really heat the atom up (thermal energy), we cause its electron to make a very large jump all the way to one of the outer shells. Smaller amounts of energy make the electron jump less far. However, as soon as it can (when we stop heating it), the electron returns to a shell closer in. Eventually it returns all the way back to shell number one. Whenever the electron jumps from an outer shell to an inner shell, it emits energy in the form of light. The energy that the electron emits is exactly the amount of energy that it absorbed when it jumped outward in the first place. Bohr discovered that all of the possible combinations of jumps that the hydrogen electron can make on its journeys back to the ground state (the first shell) equals the number of lines in the hydrogen spectrum!

This is Bohr's famous solution to the grand-piano mystery. If the electron in a hydrogen atom travels from an outer shell all the way to the innermost shell in one jump, it gives off a certain amount of energy. That makes one line in the hydrogen spectrum. If the electron in a hydrogen atom makes a tiny jump from an outer shell to the next shell inward, it gives off a much smaller amount of energy. That makes another spectral line. If the electron in a hydrogen atom jumps from shell five to shell three, for example, that makes yet another line. A jump from shell six to shell four and then from shell four to shell one makes two more spectral lines, and so on. In this way we can account for the entire hydrogen spectrum.

If we excite a hydrogen atom with white light instead of heat, we can produce the absorption phenomenon that we mentioned earlier. Each electron jump from an inner shell to a shell farther out requires a certain amount of energy, no more and no less. An electron jump from shell one to shell two requires a certain amount of energy, and only that amount. The same is true for a jump from shell five to shell seven, etc. Each jump that the electron makes from an inner shell to an outer shell takes a specific amount of energy, no more and no less.

When we shine white light on a hydrogen atom, we are offering it a whole supermarket of different energy amounts. However, it cannot use all that we have to offer; only certain specific amounts. If its electron jumps from shell one to shell four, for example, it takes that particular energy package out of the array of energy packets that we are giving it. The package that it takes out becomes a black line in the otherwise complete spectrum of white light. A jump from shell three to shell four becomes another black line. A jump from shell one to shell two, and then from shell two to shell six (there are all sorts of combinations) makes two more black lines.

In sum, if we shine white light through hydrogen gas and then through a prism, the result is the familiar white-light spectrum, but with over one hundred black lines in it. Each of these black lines corresponds to a specific energy amount that was required to make a hydrogen electron jump from one shell to another shell farther out.

These black lines in the white-light spectrum form exactly the same pattern that we get when we shine the light emitted from excited hydrogen gas directly through a prism—except, in that case, the lines are colored and the rest of the white-light spectrum is missing. Of course, the colored lines are caused by the electrons returning to lower-level shells and, in the process, emitting energy amounts equal to what they absorbed when we first made them jump. Bohr's theory permitted physicists to calculate the frequencies of the light given off by simple hydrogen atoms. These calculations agreed with observations. The grand-piano mystery was solved!

Shortly after Bohr published his theory in 1913, an army of physicists began the work of applying it to the other elements. This process was quite complicated for atoms with large numbers of electrons, and not all of the questions that physicists had about the nature of atomic phenomena were answered. Nonetheless, a tremendous amount of knowledge was gained from this work. Most of the physicists who went to work on Bohr's theory, applying it and further developing it, were *technicians*. Bohr himself, one of the founders of the new physics, was a *scientist*.

This is not to say that technicians are not important. The technician and the scientist form a partnership. Bohr could not have formulated his theory without the wealth of spectroscopic data at his disposal. That data was the result of countless laboratory hours. It was beyond Bohr's capacity, as one person, to substantiate his theory. Technicians did this for him by applying it to the other elements. Technicians are important members of the scientific community. However, since this is a book about Wu Li Masters and not about technicians, we will use the word "physicist" from now on to mean those physicists who are also scientists, that is, those physicists (people) who are not confined by the "known." From the little that we know about Wu Li Masters, it is evident that they come from this group.

There are certain limitations which no book on physics can overcome. First, there is so much to present that not even twenty volumes could contain it all. There is that much *new* material published each year. Even physicists find it impossible to keep abreast of the whole field. It requires a steady diet of reading just to keep current in one area. For everything that is included in these pages, there is much more that is not. No matter how much you learn about physics, there always will be something that is new to you. Physicists have this problem, too.

Second, no complete appreciation of physics is possible without mathematics. Nonetheless, there is no mathematics in *The Dancing Wu Li Masters*. Mathematics is a highly structured way of thinking. Physicists view the world in this way. One point of view is that they impose this structure on what they see. Another point of view is that the world presents itself most completely through such structures. In any case, mathematics is the most concise expression of physics. The reason for writing *The Dancing Wu Li Masters,* however, is that most physicists are not able to explain physics very well without it. This makes them very concise but, unfortunately, unintelligible. The fact is that most of us use words to do our explaining.

However, it is important to remember that mathematics and English are both languages. Languages are useful tools for conveying information, but if we try to communicate experiences with them, they simply do not work. All a language can do is talk *about* an experience. Wu Li Masters know that a description of an experience is not the experience. It is only talk about it.

This is a book *about* physics. Therefore, all it contains is a description. It cannot contain the experience itself. This does not mean that you will not have the experience of physics by reading it; it only means that if you do, the experience is coming from *you,* and not from the book. Quantum mechanics, for example, shows us that we are not as separate from the rest of the world as we once thought. Particle physics shows us that the "rest of the world" does not sit idly "out there." It is a sparkling realm of continual creation, transformation, and annihilation. The ideas of the new physics, when wholly grasped, can produce extraordinary *experiences*. The study of relativity theory, for example, can produce the remarkable experience that space and time are only mental constructions! Each of these different experiences is capable of changing us in such ways that we never again are able to view the world as we did before.

There is no single "experience" of physics. The experience always is changing. Relativity and quantum mechanics, although generally unknown to nonphysicists, are more than a half century old. Today, the entire field of physics is quivering with anticipation. The air is charged with excitement. A feeling is shared among physicists that radical change is at hand. A consensus grows that the near future will see new theories exploding onto the scene, incorporating the older theories and giving us a much larger view of our universe and, consequently, of ourselves.

The Wu Li Masters move in the midst of all this, now dancing this way, now that, sometimes with a heavy beat, sometimes with a lightness and grace, ever flowing freely. Now they become the dance, now the dance becomes them. This is the message of the Wu Li Masters: not to confuse the type of dance that they are doing with the fact that they are dancing.

1

※

Einstein Doesn't Like It

Quantum mechanics are not the fellows who repair automobiles in Mr. Quantum's garage. Quantum mechanics is a branch of physics. There are several branches of physics. Most physicists believe that sooner or later they will construct an overview large enough to incorporate them all.

According to this point of view, we eventually will develop, in principle, a theory which is capable of explaining everything so well that there will be nothing left to explain. This does not mean, of course, that our explanation necessarily will reflect the way that things actually are. We still will not be able to open the watch, as Einstein put it, but every occurrence in the *real* world (inside the watch) will be accounted for by a corresponding element of our final supertheory. We will have, at last, a theory that is consistent within itself and which explains all observable phenomena. Einstein called this state the "ideal limit of knowledge."[1]

This way of thinking runs into quantum mechanics the same way that the car runs into the proverbial brick wall. Einstein spent a large portion of his career arguing against quantum mechanics, even though he himself made major contributions to its development. Why did he do this? To ask this question is to stand at the edge of an abyss, still on the solid ground of Newtonian physics, but looking into the void. To answer it is to leap boldly into the new physics.

Quantum mechanics forced itself upon the scene at the beginning of this century. No convention of physicists voted to start a new branch of physics called "quantum mechanics." No one had any choice in the matter, except, perhaps, what to call it.

A "quantum" is a quantity of something, a specific amount. "Mechanics" is the study of motion. Therefore, "quantum mechanics" is the study of the motion of quantities. Quantum theory says that nature comes in bits and pieces (quanta), and quantum mechanics is the study of this phenomenon.

Quantum mechanics does not replace Newtonian physics, it includes it. The physics of Newton remains valid within its limits. To say that we have made a major new discovery about nature is one side of a coin. The other side of the coin is to say that we have found the limits of our previous theories. What we actually discover is that the way that we have been looking at nature is no longer comprehensive enough to explain all that we can observe, and we are forced to develop a more inclusive view. In Einstein's words:

> . . . creating a new theory is not like destroying an old barn and erecting a skyscraper in its place. It is rather like climbing a mountain, gaining new and wider views, discovering unexpected connections between our starting point and its rich environment. But the point from which we started out still exists and can be seen, although it appears smaller and forms a tiny part of our broad view gained by the mastery of the obstacles on our adventurous way up.[2]

Newtonian physics still is applicable to the large-scale world, but it does not work in the subatomic realm. Quantum mechanics resulted from the study of the subatomic realm, that invisible universe underlying, embedded in, and forming the fabric of everything around us.

In Newton's age (late 1600s), this realm was entirely speculation. The idea that the atom is the indivisible building block of nature was proposed about four hundred years before Christ, but until the late 1800s it remained just an idea. Then physicists developed the technology to observe the effects of atomic phenomena, thereby "proving" that atoms exist. Of course, what they really proved was that the theoretical existence of atoms was the best explanation of the experimental data that anyone could invent at the time. They also proved that atoms are not indivisible, but themselves are made of particles smaller yet, such as electrons, protons, and neutrons. These new particles were labeled "elementary particles" because physicists believed that, at last, they really had discovered the ultimate building blocks of the universe.

The elementary particle theory is a recent version of an old Greek idea. To understand the theory of elementary particles, imagine a large city made entirely of bricks. This city is filled with buildings of all shapes and sizes. Every one of them, and the streets as well, have been constructed with only a few different types of brick. If we substitute "universe" for "city" and "particle" for "brick," we have the theory of elementary particles.

It was the study of elementary particles that brought physicists nose to nose with the most devastating (to a physicist) discovery: *Newtonian physics does not work in the realm of the very small!* The impact of that earthshaking discovery still is reshaping our world view. Quantum mechanical experiments repeatedly produced results which the physics of Newton could neither predict nor explain. Yet, although Newton's physics could not account for phenomena in the microscopic realm, it continued to explain macroscopic phenomena very well (even though the macroscopic is made of the microscopic)! This was perhaps the most profound discovery of science.

Newton's laws are based upon observations of the everyday world. They predict events. These events pertain to real things like baseballs and bicycles. Quantum mechanics is based upon experiments conducted in the subatomic realm. It predicts probabilities. These probabilities pertain to subatomic phenomena. Subatomic phenom-

ena cannot be observed directly. None of our senses can detect them.*
Not only has no one ever seen an atom (much less an electron), no
one has ever tasted, touched, heard, or smelled one either.

Newton's laws depict events which are simple to understand and
easy to picture. Quantum mechanics depicts the probabilities of phenomena which defy conceptualization and are impossible to visualize.
Therefore, these phenomena must be understood in a way that is not
more difficult than our usual way of understanding, but different from
it. Do not try to make a complete mental picture of quantum mechanical events. (Physicists make partial pictures of quantum phenomena,
but even these pictures have a questionable value.) Instead, allow
yourself to be open without making an effort to visualize anything.
Werner Heisenberg, one of the founders of quantum physics, wrote:

> The mathematically formulated laws of quantum theory show
> clearly that our ordinary intuitive concepts cannot be unambig-
> uously applied to the smallest particles. All the words or con-
> cepts we use to describe ordinary physical objects, such as
> position, velocity, color, size, and so on, become indefinite and
> problematic if we try to use them of elementary particles.[3]

The idea that we do not understand something until we have a
picture of it in our heads is a by-product of the Newtonian way of
looking at the world. If we want to get past Newton, we have to get
past that.

Newton's first great contribution to science was the laws of motion.
If an object, said Newton, is moving in a straight line, it will continue
moving in a straight line forever unless it is acted upon by something
else (a "force"). At that time its direction and speed will be altered,
depending upon the magnitude and direction of the force which it

---

*The dark-adapted eye can detect a single photon. Otherwise, only the
effects of subatomic phenomena are available to our senses (a track on a photographic plate, a pointer movement on a meter, etc.).

encounters. Furthermore, every action is accompanied by an equal and opposite reaction.

Today, these concepts are familiar to anyone who has studied physics or hung out in a pool hall. However, if we mentally project ourselves three hundred years into the past, we can see how remarkable they really are.

First, Newton's first law of motion defied the accepted authority of the day, which was Aristotle. According to Aristotle, the natural inclination for a moving object is to return to a state of rest.

Second, Newton's laws of motion describe events which were unobservable in the 1600s. In the everyday world, which was all that Newton had to observe, moving objects always *do* return to a state of rest because of friction. If we put a wagon in motion, it encounters friction from the air through which it passes, from the ground its tires move on, from the axles that its wheels turn around, and, unless it is rolling downhill, sooner or later it comes to rest. We can streamline the wagon, grease the wheels, and use a smooth road, but this only reduces the effect of friction. Eventually the wagon stops moving, apparently on its own.

Newton never had the chance to see a film of astronauts in space, but he predicted what they would encounter. When an astronaut releases a pencil in front of him, nothing happens. It just stays there. If he gives it a push, off it goes in the direction of the push until it bumps into a wall. If the wall were not there, the pencil would continue to move uniformly, in principle, forever. (The astronaut also moves off in the opposite direction, but much more slowly because of his greater mass.)

Third, Newton's premise was "I make no hypotheses" ("Hypotheses non fingo"), which means that he based his laws upon sound experimental evidence, and nothing else. His criteria for the validity of everything that he wrote was that anyone should be able to reproduce his experiments and come up with the same results. If it could be verified experimentally, it was true. If it could not be verified experimentally, it was suspect.

The church took a dim view, to say the least, of this position.

Since it had been saying things for fifteen hundred years which hardly were subject to experimental verification, Newtonian physics, in effect, was a direct challenge to the power of the church. The power of the church was considerable.* Shortly before Newton's birth, Galileo was seized by the Inquisition for declaring that the earth revolves around the sun and for drawing unacceptable theological implications from his beliefs. He was forced to recant on penalty of imprisonment or worse. This made a considerable impression on many people, among them another founder of modern science, the Frenchman René Descartes.

In the 1630s Descartes visited the royal gardens at Versailles, which were known for their intricate automata. When water was made to flow, music sounded, sea nymphs began to play, and a giant Neptune, complete with trident, advanced menacingly. Whether the idea was in his mind before this visit or not, Descartes's philosophy, which he supported with his mathematics, became that the universe and all of the things in it also were automata. From Descartes's time to the beginning of this century, and perhaps because of him, our ancestors began to see the universe as a Great Machine. Over the next three hundred years they developed science specifically to discover how the Great Machine worked.

Newton's second great contribution to science was his law of gravity. Gravity is a remarkable phenomenon, even though we take it for granted. For example, if we hold a ball off the ground, and then release it, the ball falls straight down to the ground. But how did that happen? The ground did not reach up and pull the ball down, yet the ball was pulled to the earth. The old physics called this unexplainable phenomenon "action-at-a-distance." Newton himself was as puzzled as anyone. He wrote in his famous *Philosophiae Naturalis Principia Mathematica:*

*At the time of Newton's discoveries, the power of the church already had been challenged by Martin Luther. Newton himself was a pious person. The specific argument of the church was not with empirical method, but with the theological conclusions that were being developed from Newton's ideas, conclusions which involved the concept of God as creator and the central position of man in creation.

. . . I have not been able to discover the cause of those properties of gravity from phenomena, and I frame no hypotheses . . . it is enough that gravity does really exist, and act according to the laws which we have explained, and abundantly serves to account for all the motions of the celestial bodies . . .[4]

Newton clearly felt that a true understanding of the nature of gravity was beyond comprehension. In a letter to Richard Bently, a classical scholar, he wrote:

. . . that one body may act upon another at a distance through a vacuum without the mediation of anything else, by and through which their action and force may be conveyed from one to another, is to me so great an absurdity that, I believe, no man who has in philosophic matters a competent faculty of thinking could ever fall into it.[5]

In short, action-at-a-distance could be described, but it could not be explained.

Newton's thesis was that the same force which pulls apples downward also keeps the moon in orbit around the earth and the planets in orbit around the sun. To test his idea, he calculated various movements of the moon and the planets, using his own mathematics. Then he compared his findings with the observations of astronomers. His calculations and their observations matched! In one stroke Newton set aside the assumption of an essential difference between earthly and heavenly objects by showing that both of them are governed by the same laws. He established a rational celestial mechanics. What had been the purview of the gods, or God, came now within the comprehension of mortals. Newton's gravitational law does not explain gravity (that was done by Einstein in his general theory of relativity) but it does subject the effects of gravity to a rigorous mathematical formalism.

Newton was the first person to discover principles in nature which unify large tracts of experience. He abstracted certain unifying concepts from the endless diversity of nature and gave those concepts mathematical expression. Because of this, more than anything else, Newton's work has influenced us so forcefully. Newton showed us that the phenomena of the universe are structured in rationally comprehensible ways. He gave us the most powerful tool in history. In the West we have used this tool, if not wisely, certainly to the best of our ability. The results, both positive and negative, have been spectacular. The story of our enormous impact on our environment begins with the work of Newton.

It was Galileo Galilei who, following the Middle Ages, first quantified the physical world. He measured the motion, frequency, velocity, and duration of everything from falling stones to swinging pendulums (like the chandelier in his cathedral). It was René Descartes who developed many of the fundamental techniques of modern mathematics and gave us the picture of the universe as a Great Machine. It was Isaac Newton who formulated the laws by which the Great Machine runs.

These men struck boldly against the grip of scholasticism, the medieval thought system of the 12th to the 15th centuries. They attempted to place "man" at the center of the stage, or at least back on the stage; to prove to him that he need not be a bystander in a world governed by unfathomable forces. It is perhaps the greatest irony of history that they accomplished just the opposite.

Joseph Weizenbaum, a scientist at the Massachusetts Institute of Technology, wrote, in reference to computers:

> Science promised man power. . . . But, as so often happens when people are seduced by promises of power, the price is servitude and impotence. Power is nothing if it is not the power to choose.[6]

How did this happen?

Newton's laws of motion describe what happens to a moving object. Once we know the laws of motion we can predict the future of a moving object provided that we know certain things about it initially. The more initial information that we have, the more accurate our predictions will be. We also can retrodict (predict backward in time) the past history of a given object. For example, if we know the present position and velocity of the earth, the moon, and the sun, we can predict where the earth will be in relation to the moon and the sun at any particular time in the future, giving us a foreknowledge of eclipses, seasons, and so on. In like manner, we can calculate where the earth has been in relation to the moon and the sun, and when similar phenomena occurred in the past.

Without Newtonian physics the space program would not be possible. Moon probes are launched at the precise moment when the launch site on the earth (which simultaneously is rotating around its axis and moving forward through space) is in a position, relative to the landing zone on the moon (which also is rotating and moving) such that the path traversed by the spacecraft is the shortest possible. The calculations of the earth, moon, and spacecraft movements are done by computer, but the mechanics used are the same ones that are described in Newton's *Philosophiae Naturalis Principia Mathematica*.

In practice, it is very difficult to know all the initial circumstances pertaining to an event. Even a simple action such as bouncing a ball off a wall is surprisingly complex. The shape, size, elasticity, and momentum of the ball, the angle at which it was thrown, the density, pressure, humidity and temperature of the air, the shape, hardness, and position of the wall, to name a few of the essential elements, are all required to know where and when the ball will land. It is increasingly difficult to obtain all of the data necessary for accurate predictions when more complex actions are involved. According to the old physics, however, it is possible, in principle, to predict *exactly* how a given event is going to unfold if we have enough information about it. In practice, it is only the enormity of the task that prevents us from accomplishing it.

The ability to predict the future based on a knowledge of the present and the laws of motion gave our ancestors a power they had never known. However, these concepts carry within them a very dispiriting logic. If the laws of nature determine the future of an event, then, given enough information, we could have predicted our present at some time in the past. That time in the past also could have been predicted at a time still earlier. In short, if we are to accept the mechanistic determination of Newtonian physics—if the universe really is a great machine—then from the moment that the universe was created and set into motion, everything that was to happen in it already was determined.

According to this philosophy, we may seem to have a will of our own and the ability to alter the course of events in our lives, but we do not. Everything, from the beginning of time, has been predetermined, including our illusion of having a free will. The universe is a prerecorded tape playing itself out in the only way that it can. The status of men is immeasurably more dismal than it was before the advent of science. The Great Machine runs blindly on, and all things in it are but cogs.

According to quantum mechanics, however, it is not possible, *even in principle,* to know enough about the present to make a complete prediction about the future. Even if we have the best possible measuring devices, it is not possible. It is not a matter of the size of the task or the inefficiency of detectors. The very nature of things is such that we must choose which aspect of them we wish to know best, for we can know only one of them with precision.

As Niels Bohr, another founder of quantum mechanics, put it:

> . . . in quantum mechanics, we are not dealing with an arbitrary renunciation of a more detailed analysis of atomic phenomena, but with a recognition that such an analysis is *in principle* excluded.[7] [Italics in the original]

For example, imagine an object moving through space. It has both a position and a momentum which we can measure. This is an example of the old (Newtonian) physics. (Momentum is a combination of how big an object is, how fast it is going, and the direction that it is moving.) Since we can determine both the position and the momentum of the object at a particular time, it is not a very difficult affair to calculate where it will be at some point in the future. If we see an airplane flying north at two hundred miles per hour, we know that in one hour it will be two hundred miles farther north if it does not change its course or speed.

The mind-expanding discovery of quantum mechanics is that Newtonian physics does not apply to subatomic phenomena. In the subatomic realm, we cannot know both the position *and* the momentum of a particle with absolute precision. We can know both, approximately, but the more we know about one, the less we know about the other. We can know either of them precisely, but in that case, we can know nothing about the other. This is Werner Heisenberg's uncertainty principle. As incredible as it seems, it has been verified repeatedly by experiment.

Of course, if we picture a moving particle, it is very difficult to imagine not being able to measure both its position and momentum. Not to be able to do so defies our "common sense." This is not the only quantum mechanical phenomenon which contradicts common sense. Commonsense contradictions, in fact, are at the heart of the new physics. They tell us again and again that the world may not be what we think it is. It may be much, much more.

Since we cannot determine both the position and momentum of subatomic particles, we cannot predict much about them. Accordingly, quantum mechanics does not and cannot predict specific events. It does, however, predict *probabilities*. Probabilities are the odds that something is going to happen, or that it is not going to happen. Quantum theory can predict the probability of a microscopic event with the same precision that Newtonian physics can predict the actual occurrence of a macroscopic event.

Newtonian physics says, "If such and such is the case now, then such and such is going to happen next." Quantum mechanics says, "If such and such is the case now, then the *probability* that such and such is going to happen next is . . . (whatever it is calculated to be)." We never can know with certainty what will happen to the particle that we are "observing." All that we can know for sure are the probabilities for it to behave in certain ways. This is the most that we can know because the two data which must be included in a Newtonian calculation, position and momentum, cannot both be known with precision. *We must choose,* by the selection of our experiment, which one we want to measure most accurately.

The lesson of Newtonian physics is that the universe is governed by laws that are susceptible to rational understanding. By applying these laws we extend our knowledge of, and therefore our influence over, our environment. Newton was a religious person. He saw his laws as manifestations of God's perfection. Nonetheless, Newton's laws served man's cause well. They enhanced his dignity and vindicated his importance in the universe. Following the Middle Ages, the new field of science ("Natural Philosophy") came like a fresh breeze to revitalize the spirit. It is ironic that, in the end, Natural Philosophy reduced the status of men to that of helpless cogs in a machine whose functioning had been preordained from the day of its creation.

Contrary to Newtonian physics, quantum mechanics tells us that our knowledge of what governs events on the subatomic level is not nearly what we assumed it would be. It tells us that we cannot predict subatomic phenomena with any certainty. We only can predict their probabilities.

Philosophically, however, the implications of quantum mechanics are psychedelic. Not only do we influence our reality, but, in some degree, we actually *create* it. Because it is the nature of things that we can know either the momentum of a particle or its position, but not both, *we must choose* which of these two properties we want to determine. Metaphysically, this is very close to saying that we *create* certain properties because we choose to measure those properties. Said another way, it is possible that we create something that has position, for example, like a particle, because we are intent on determining position and it is impossible to determine position without having some *thing* occupying the position that we want to determine.

Quantum physicists ponder questions like, "Did a particle with momentum exist before we conducted an experiment to measure its momentum?"; "Did a particle with position exist before we conducted an experiment to measure its position?"; and "Did any particles exist at all before we thought about them and measured them?"

"Did we create the particles that we are experimenting with?" Incredible as it sounds, this is a possibility that many physicists recognize.

John Wheeler, a well-known physicist at Princeton, wrote:

> May the universe in some strange sense be "brought into being" by the participation of those who participate? . . . The vital act is the act of participation. "Participator" is the incontrovertible new concept given by quantum mechanics. It strikes down the term "observer" of classical theory, the man who stands safely behind the thick glass wall and watches what goes on without taking part. It can't be done, quantum mechanics says.[8]

The languages of eastern mystics and western physicists are becoming very similar.

Newtonian physics and quantum mechanics are partners in a double irony. Newtonian physics is based upon the idea of laws which govern phenomena and the power inherent in understanding them, but it leads to impotence in the face of a Great Machine which is the universe. Quantum mechanics is based upon the idea of minimal knowledge of future phenomena (we are limited to knowing probabilities) but it leads to the possibility that our reality is what we choose to make it.

There is another fundamental difference between the old physics and the new physics. The old physics assumes that there is an external world which exists apart from us. It further assumes that we can observe, measure, and speculate about the external world without changing it. According to the old physics, the external world is indifferent to us and to our needs.

Galileo's historical stature stems from his tireless (and successful) efforts to quantify (measure) the phenomena of the external world. There is great power inherent in the process of quantification. For example, once a relationship is discovered, like the rate of acceleration of a falling object, it matters not who drops the object, what object is dropped, or where the dropping takes place. The results are always the same. An experimenter in Italy gets the same results as a Russian experimenter who repeats the experiment a century later. The results are the same whether the experiment is done by a skeptic, a believer, or a curious bystander.

Facts like these convinced philosophers that the physical universe goes unheedingly on its way, doing what it must, without regard for its inhabitants. For example, if we simultaneously drop two people from the same height, it is a verifiable (repeatable) fact that they both will hit the ground at the same time, regardless of their weights. We can measure their fall, acceleration, and impact the same way that we measure the fall, acceleration, and impact of stones. In fact, the results will be the same as if they *were* stones.

"But there is a difference between people and stones!" you might say. "Stones have no opinions or emotions. People have both. One of these dropped people, for example, might be frightened by his experience and the other might be angry. Don't their feelings have any importance in this scheme?"

No. The feelings of our subjects matter not in the least. When we take them up the tower again (struggling this time) and drop them off again, they fall with the same acceleration and duration that they did the first time, even though now, of course, they are both fighting mad. The Great Machine is impersonal. In fact, it was precisely this impersonality that inspired scientists to strive for "absolute objectivity."

The concept of scientific objectivity rests upon the assumption of an external world which is "out there" as opposed to an "I" which is "in here." (This way of perceiving, which puts other people "out there," makes it very lonely "in here.") According to this view, Nature, in all her diversity, is "out there." The task of the scientist is to observe the "out there" as objectively as possible. To observe something objectively means to see it as it would appear to an observer who has no prejudices about what he observes.

The problem that went unnoticed for three centuries is that a person who carries such an attitude certainly is prejudiced. His prejudice is to be "objective," that is, to be without a preformed opinion. In fact, it is impossible to be without an opinion. An opinion is a point of view. The point of view that we can be without a point of view is a point of view. The decision itself to study one segment of reality instead of another is a subjective expression of the researcher who makes it. It affects his perceptions of reality, if nothing else. Since reality is what we are studying, the matter gets very sticky here.

The new physics, quantum mechanics, tells us clearly that it is not possible to observe reality without changing it. If we observe a certain particle collision experiment, not only do we have no way of proving that the result would have been the same if we had not been watching it, all that we know indicates that it would not have been the same, because the result that we got was affected by the fact that we were looking for it.

Some experiments show that light is wave-like. Other experiments show equally well that light is particle-like. If we want to demonstrate that light is a particle-like phenomenon or that light is a wave-like phenomenon, we only need to select the appropriate experiment.

According to quantum mechanics there is no such thing as objectivity. We cannot eliminate ourselves from the picture. We are a part of nature, and when we study nature there is no way around the fact that nature is studying itself. Physics has become a branch of psychology, or perhaps the other way round.

Carl Jung, the Swiss psychologist, wrote:

> The psychological rule says that when an inner situation is not made conscious, it happens outside, as fate. That is to say, when the individual remains undivided and does not become conscious of his inner contradictions, the world must perforce act out the conflict and be torn into opposite halves.[9]

Jung's friend, the Nobel Prize–winning physicist, Wolfgang Pauli, put it this way:

> From an inner center the psyche seems to move outward, in the sense of an extraversion, into the physical world . . .[10]

If these men are correct, then physics is the study of the structure of consciousness.

The descent downward from the macroscopic level to the microscopic level, which we have been calling the realm of the very small, is a two-step process. The first step downward is to the atomic level. The second step downward is to the subatomic level.

The smallest object that we can see, even under a microscope, contains millions of atoms. To see the atoms in a baseball, we would have to make the baseball the size of the earth. If a baseball were the size of the earth, its atoms would be about the size of grapes. If you can picture the earth as a huge glass ball filled with grapes, that is approximately how a baseball full of atoms would look.

The step downward from the atomic level takes us to the subatomic level. Here we find the particles that make up atoms. The difference between the atomic level and the subatomic level is as great as the difference between the atomic level and the world of sticks and rocks. It would be impossible to see the nucleus of an atom the size of a grape. In fact, it would be impossible to see the nucleus of an atom the size of a room. To see the nucleus of an atom, the atom would have to be as high as a fourteen-story building! The nucleus of an atom as high as a fourteen-story building would be about the size of a grain of salt. Since a nuclear particle has about 2,000 times more mass than an electron, the electrons revolving around this nucleus would be about as massive as dust particles!

The dome of Saint Peter's basilica in the Vatican has a diameter of about fourteen stories. Imagine a grain of salt in the middle of the dome of Saint Peter's with a few dust particles revolving around it at the outer edges of the dome. This gives us the scale of subatomic particles. It is in this realm, the subatomic realm, that Newtonian physics has proven inadequate, and that quantum mechanics is required to explain particle behavior.

A subatomic particle is not a "particle" like a dust particle. There is more than a difference in size between a dust particle and a subatomic particle. A dust particle is a *thing*, an object. A subatomic particle cannot be pictured as a thing. Therefore, we must abandon the idea of a subatomic particle as an object.

Quantum mechanics views subatomic particles as "tendencies to exist" or "tendencies to happen." How strong these tendencies are is expressed in terms of probabilities. A subatomic particle is a "quantum," which means a quantity of something. What that something is, however, is a matter of speculation. Many physicists feel that it is not meaningful even to pose the question. It may be that the search for the ultimate "stuff" of the universe is a crusade for an illusion. At the subatomic level, mass and energy change unceasingly into each other. Particle physicists are so familiar with the phenomena of mass becoming energy and energy becoming mass that they routinely measure the mass of particles in energy units.* Since the tendencies of subatomic phenomena to become manifest under certain conditions are probabilities, this brings us to the matter (no pun) of statistics.

Because there are millions of millions of subatomic particles in the smallest space that we can see, it is convenient to deal with them statistically. Statistical descriptions are pictures of crowd behavior. Statistics cannot tell us how one individual in a crowd will behave, but they can give us a fairly accurate description, based on repeated observations, of how a group as a whole behaves.

For example, a statistical study of population growth may tell us how many children were born in each of several years and how many are predicted to be born in years to come. However, the statistics cannot tell us which families will have the new children and which ones will not. If we want to know the behavior of traffic at an intersection, we can install devices there to gather data. The statistics that these devices provide may tell us how many cars, for instance, turn left during certain hours, but not *which* cars.

---

*Strictly speaking, mass, according to Einstein's special theory of relativity, *is* energy and energy *is* mass. Where there is one, there is the other.

Statistics is used in Newtonian physics. It is used, for example, to explain the relationship between gas volume and pressure. This relation is named Boyle's Law after its discoverer, Robert Boyle, who lived in Newton's time. It could as easily be known as the Bicycle Pump Law, as we shall see. Boyle's Law says that if the volume of a container holding a given amount of gas at a constant temperature is reduced by one half, the pressure exerted by the gas in the container doubles.

Imagine a person with a bicycle pump. He has pulled the plunger fully upward, and is about ready to push it down. The hose of the pump is connected to a pressure gauge instead of to a bicycle tire, so that we can see how much pressure is in the pump. Since there is no pressure on the plunger, there is no pressure in the pump cylinder and the gauge reads zero. However, the pressure inside the pump is not actually zero. We live at the bottom of an ocean of air (our atmosphere). The weight of the several miles of air above us exerts a pressure at sea level of 14.7 pounds on every square inch of our bodies. Our bodies do not collapse because they are exerting 14.7 pounds per square inch outward. This is the state that we usually read as zero on a bicycle pressure gauge. To be accurate, suppose that we set our gauge to read 14.7 pounds per square inch before we push down on the pump handle.

Now we push the piston down halfway. The interior volume of the pump cylinder is now one half of its original size, and no air has been allowed to escape, because the hose is connected to a pressure gauge. The gauge now reads 29.4 pounds per square inch, or twice the original pressure. Next we push the plunger two thirds of the way down. The interior volume of the pump cylinder is now one third of its original size, and the pressure gauge reads three times the original pressure (44.1 pounds per square inch). This is Boyle's Law: At a constant temperature the pressure of a quantity of gas is inversely proportional to its volume. If the volume is reduced to one half, the pressure doubles; if the volume is reduced to one third, the pressure triples, etc. To explain why this is so, we come to classical statistics.

The air (a gas) in our pump is composed of millions of molecules (molecules are made of atoms). These molecules are in constant motion, and at any given time, millions of them are banging into the pump walls. Although we do not detect each single collision, the macroscopic effect of these millions of impacts on a square inch of the pump wall produces the phenomenon of "pressure" on it. If we reduce the volume of the pump cylinder by one half, we crowd the gas molecules into a space twice as small as the original one, thereby causing twice as many impacts on the same square inch of pump wall. The macroscopic effect of this is a doubling of the "pressure." By crowding the molecules into one third of the original space, we cause three times as many molecules to bang into the same square inch of pump wall, and the "pressure" on it triples. This is the kinetic theory of gases.

In other words, "pressure" results from the group behavior of a large number of molecules in motion. It is a collection of individual events. Each individual event can be analyzed because, according to Newtonian physics, each individual event is theoretically subject to deterministic laws. In principle, we can calculate the path of each molecule in the pump chamber. This is how statistics is used in the old physics.

Quantum mechanics also used statistics, but there is a very big difference between quantum mechanics and Newtonian physics. In quantum mechanics, there is no way to predict individual events. This is the startling lesson that experiments in the subatomic realm have taught us.

Therefore, quantum mechanics concerns itself only with group behavior. It intentionally leaves vague the relation between group behavior and individual events because individual subatomic events cannot be determined accurately (the uncertainty principle) and, as we shall see in high-energy particles, they constantly are changing. Quantum physics abandons the laws which govern individual events and states *directly* the statistical laws which govern collections of events. Quantum mechanics can tell us how a group of particles will behave, but the only thing that it can say about an individual particle is how it *probably* will behave. Probability is one of the major characteristics of quantum mechanics.

This makes quantum mechanics an ideal tool for dealing with subatomic phenomena. For example, take the phenomenon of common radioactive decay (luminous watch dials). Radioactive decay is a phenomenon of predictable overall behavior consisting of unpredictable individual events.

Suppose that we put one gram of radium in a time vault and leave it there for sixteen hundred years. When we return, do we find one gram of radium? No! We find only half a gram. This is because radium atoms naturally disintegrate at a rate such that every sixteen hundred years half of them are gone. Therefore, physicists say that radium has a "half life" of sixteen hundred years. If we put the radium back in the vault for another sixteen hundred years, only one fourth of the original gram would remain when we opened the vault again. Every sixteen hundred years one half of all the radium atoms in the world disappear. How do we know which radium atoms are going to disintegrate and which radium atoms are not going to disintegrate?

We don't. We can predict how many atoms in a piece of radium are going to disintegrate in the next hour, but we have no way of determining *which* ones are going to disintegrate. There is no physical law that we know of which governs this selection. Which atoms decay is purely a matter of chance. Nonetheless, radium continues to decay, on schedule, as it were, with a precise and unvarying half life of sixteen hundred years. Quantum theory dispenses with the laws governing the disintegration of individual radium atoms and proceeds directly to the statistical laws governing the disintegration of radium atoms as a group. This is how statistics is used in the new physics.

Another good example of predictable overall (statistical) behavior consisting of unpredictable individual events is the constant variation of intensity among spectral lines. Remember that, according to Bohr's theory, the electrons of an atom are located only in shells which are specific distances from the nucleus (page 14). Normally, the single electron of a hydrogen atom remains in the shell closest to the nucleus (the ground state). If we excite it (add energy to it) we cause it to jump to a shell farther out. The more energy we give it, the farther out it jumps. If we stop exciting it, the electron jumps inward to a shell closer to the nucleus, eventually returning all the way to the innermost shell. With each jump from an outer shell to an inner shell, the electron emits an energy amount equal to the energy amount that it absorbed when we caused it to jump outward. These emitted energy packets (photons) constitute the light which, when dispersed through a prism, forms the spectrum of one hundred or so colored lines that is peculiar to hydrogen. Each colored line in the hydrogen spectrum is made from the light emitted from hydrogen electrons as they jump from a particular outer shell to a particular inner shell.

What we did not mention earlier is that some of the lines in the hydrogen spectrum are more pronounced than others. The lines that are more pronounced are always more pronounced and the lines that are faint are always faint. The intensity of the lines in the hydrogen spectrum varies because hydrogen electrons returning to the ground state do not always take the same route.

Shell five, for example, may be a more popular stopover than shell three. In that case, the spectrum produced by millions of excited hydrogen atoms will show a more pronounced spectral line corresponding to electron jumps from shell five to shell one and a less pronounced spectral line corresponding to electron jumps from, say, shell three to shell one. That is because, in this example, more electrons stop over at shell five before jumping to shell one than stop over at shell three before jumping to shell one.

In other words, the probability is very high, in this example, that the electrons of excited hydrogen atoms will stop at shell five on their way back to shell one, and the probability is lower that they will stop at shell three. Said another way, we know that a certain number of electrons probably will stop at shell five and that a certain lesser number of electrons probably will stop at shell three. Still, we have no way of knowing *which* electrons will stop where. As before, we can describe precisely an overall behavior without being able to predict a single one of the individual events which comprise it.

*   *   *

This brings us to the central philosophical issue of quantum mechanics, namely, "What is *it* that quantum mechanics describes?" Put another way, quantum mechanics statistically describes the overall behavior and/or predicts the probabilities of the individual behavior of what?

In the autumn of 1927, physicists working with the new physics met in Brussels, Belgium, to ask themselves this question, among others. What they decided there became known as the Copenhagen Interpretation of Quantum Mechanics.* Other interpretations developed later, but the Copenhagen Interpretation marks the emergence of the new physics as a consistent way of viewing the world. It is still the most prevalent interpretation of the mathematical formalism of quantum mechanics. The upheaval in physics following the discovery of the inadequacies of Newtonian physics was all but complete. The question among the physicists at Brussels was not whether Newtonian mechanics could be adapted to subatomic phenomena (it was clear that it could not be), but rather, what was to replace it.

The Copenhagen Interpretation was the first consistent formulation of quantum mechanics. Einstein opposed it in 1927 and he argued against it until his death, although he, like all physicists, was forced to acknowledge its advantages in explaining subatomic phenomena.

The Copenhagen Interpretation says, in effect, that *it does not matter* what quantum mechanics is about!† The important thing is that it works in all possible experimental situations. This is one of the most important statements in the history of science. The Copenhagen Interpretation of Quantum Mechanics began a monumental reunion which was all but unnoticed at the time. The rational part of our psyche, typified by science, began to merge again with that other part of us which we had ignored since the 1700s, our irrational side.

---

*This was the 5th Solvay Congress at which Bohr and Einstein conducted their now famous debates. The term "Copenhagen Interpretation" reflects the dominant influence of Niels Bohr (from Copenhagen) and his school of thought.

†The Copenhagen Interpretation says that quantum theory is about correlations in our experiences. It is about what will be observed under specified conditions.

The scientific idea of truth traditionally had been anchored in an absolute truth somewhere "out there"—that is, an absolute truth with an independent existence. The closer that we came in our approximations to the absolute truth, the truer our theories were said to be. Although we might never be able to perceive the absolute truth directly—or to open the watch, as Einstein put it—still we tried to construct theories such that for every facet of absolute truth, there was a corresponding element in our theories.

The Copenhagen Interpretation does away with this idea of a one-to-one correspondence between reality and theory. This is another way of saying what we have said before. Quantum mechanics discards the laws governing individual events and states directly the laws governing aggregations. It is very pragmatic.

The philosophy of pragmatism goes something like this. The mind is such that it deals only with ideas. It is not possible for the mind to relate to anything other than ideas. Therefore, it is not correct to think that the mind actually can ponder reality. All that the mind can ponder is its *ideas* about reality. (Whether or not that is the way reality actually is, is a metaphysical issue.) Therefore, whether or not something is true is not a matter of how closely it corresponds to the absolute truth, but of how consistent it is with our experience.*

The extraordinary importance of the Copenhagen Interpretation lies in the fact that for the first time, scientists attempting to formulate

*The philosophy of pragmatism was created by the American psychologist William James. Recently, the pragmatic aspects of the Copenhagen Interpretation of Quantum Mechanics have been emphasized by Henry Pierce Stapp, a theoretical physicist at the Lawrence Berkeley Laboratory in Berkeley, California. The Copenhagen Interpretation, in addition to the pragmatic part, has the claim that quantum theory is in some sense complete; that no theory can explain subatomic phenomena in any more detail.

An essential feature of the Copenhagen Interpretation is Bohr's principle of complementarity (to be discussed later). Some historians practically equate the Copenhagen Interpretation and complementarity. Complementarity is subsumed in a general way in Stapp's pragmatic interpretation of quantum mechanics, but the special emphasis on complementarity is characteristic of the Copenhagen Interpretation.

a consistent physics were forced by their own findings to acknowledge that a complete understanding of reality lies beyond the capabilities of rational thought. It was this that Einstein could not accept. "The most incomprehensible thing about the world," he wrote, "is that it is comprehensible."[11] But the deed was done. The new physics was based not upon "absolute truth," but upon *us*.

Henry Pierce Stapp, a physicist at the Lawrence Berkeley Laboratory, expressed this eloquently:

> [The Copenhagen Interpretation of Quantum Mechanics] was essentially a rejection of the presumption that nature could be understood in terms of elementary space-time realities. According to the new view, the complete description of nature at the atomic level was given by probability functions that referred, not to underlying microscopic space-time realities, but rather to the macroscopic objects of sense experience. The theoretical structure did not extend down and anchor itself on fundamental microscopic space-time realities. Instead it turned back and anchored itself in the concrete sense realities that form the basis of social life. . . . This pragmatic description is to be contrasted with descriptions that attempt to peer "behind the scenes" and tell us what is "really happening."[12]

Another way of understanding the Copenhagen Interpretation (in retrospect) is in terms of split-brain analysis. The human brain is divided into two halves which are connected at the center of the cerebral cavity by a tissue. To treat certain conditions, such as epilepsy, the two halves of the brain sometimes are separated surgically. From the experiences reported by and the observations made of persons who have undergone this surgery, we have discovered a remarkable fact. Generally speaking, the left side of our brain functions in a different manner than the right side. Each of our two brains sees the world in a different way.

The left side of our brain perceives the world in a linear manner. It tends to organize sensory input into the form of points on a line, with some points coming before others. For example, language, which is linear (the words which you are reading flow along a line from left to right), is a function of the left hemisphere. The left hemisphere functions logically and rationally. It is the left side of the brain which creates the concept of causality, the image that one thing causes another because it always precedes it. The right hemisphere, by comparison, perceives whole patterns.

Persons who have had split-brain operations actually have two separate brains. When each hemisphere is tested separately, it is found that the left brain remembers how to speak and use words, while the right brain generally cannot. However, the right brain remembers the lyrics of songs! The left side of our brain tends to ask certain questions of its sensory input. The right side of our brain tends to accept what it is given more freely. Roughly speaking, the left hemisphere is "rational" and the right hemisphere is "irrational."[13]

Physiologically, the left hemisphere controls the right side of the body and the right hemisphere controls the left side of the body. In view of this, it is no coincidence that both literature and mythology associate the right hand (left hemisphere) with rational, male, and assertive characteristics and the left hand (right hemisphere) with mystical, female, and receptive characteristics. The Chinese wrote about the same phenomena thousands of years ago (yin and yang) although they were not known for their split-brain surgery.

Our entire society reflects a left hemispheric bias (it is rational, masculine, and assertive). It gives very little reinforcement to those characteristics representative of the right hemisphere (intuitive, feminine, and receptive). The advent of "science" marks the beginning of the ascent of left hemispheric thinking into the dominant mode of western cognition and the descent of right hemispheric thinking into the underground (underpsyche) status from which it did not emerge (with scientific recognition) until Freud's discovery of the "unconscious" which, of course, he labeled dark, mysterious, and irrational (because that is how the left hemisphere views the right hemisphere).

The Copenhagen Interpretation was, in effect, a recognition of the limitations of left hemispheric thought, although the physicists at Brussels in 1927 could not have thought in those terms. It was also a *re-cognition* of those psychic aspects which long had been ignored in a rationalistic society. After all, physicists are essentially people who wonder at the universe. To stand in awe and wonder is to understand in a very specific way, even if that understanding cannot be described. The subjective experience of wonder is a message to the rational mind that the object of wonder is being perceived and understood in ways other than the rational.

The next time you are awed by something, let the feeling flow freely through you and do not try to "understand" it. You will find that you *do understand*, but in a way that you will not be able to put into words. You are perceiving intuitively through your right hemisphere. It has not atrophied from lack of use, but our skill in listening to it has been dulled by three centuries of neglect.

Wu Li Masters perceive in both ways, the rational and the irrational, the assertive and the receptive, the masculine and the feminine. They reject neither one nor the other. They only dance.

DANCING LESSON FOR NEWTONIAN PHYSICS	DANCING LESSON FOR QUANTUM MECHANICS
Can picture it.	Cannot picture it.
Based on ordinary sense perceptions.	Based on behavior of subatomic particles and systems not directly observable.
Describes *things*; individual objects in space and their changes in time.	Describes statistical behavior of *systems*.
Predicts events.	Predicts probabilities.
Assumes an objective reality "out there."	Does not assume an objective reality apart from our experience.
We can observe something without changing it.	We cannot observe something without changing it.

DANCING LESSON FOR NEWTONIAN PHYSICS	DANCING LESSON FOR QUANTUM MECHANICS
Claims to be based on "absolute truth"; the way that nature really is "behind the scenes."	Claims only to correlate experience correctly.

This is quantum mechanics. The next question is, "How does it work?"

Part One

PATTERNS OF ORGANIC ENERGY

1

✺

Living?

When we talk of physics as patterns of organic energy, the word that catches our attention is "organic." Organic means living. Most people think that physics is about things that are not living, such as pendulums and billiard balls. This is a common point of view, even among physicists, but it is not as evident as it may seem.

Let us explore this viewpoint with the aid of a hypothetical person, a young man named Jim de Wit, who is the perpetual champion of the non-obvious.

"It is not at all true," says Jim de Wit, "that physics is about nonliving things. This is evident from our discussion of falling bodies (page 32). Even if some of them are the human kind, they all accelerate at the same rate in a vacuum. So physics does apply to living things."

"But that is an unfair example," we say. "Rocks have no choice in the matter of falling. If we drop them, they fall. If we don't drop them, they don't fall. Humans, on the other hand, exercise choice. Accidents excluded, humans ordinarily are not found in the act of falling. Why? Because they know that falling may hurt them and they have no desire to be hurt. In other words, humans process *informa-*

tion (they know that they may be hurt) and they *respond* to it (by not falling). Rocks can do neither."

"That is the way things appear," says de Wit, "but it may not be the way they actually are. For example, by watching time-lapse photography we know that plants often respond to stimulae with humanlike reactions. They retreat from pain, advance toward pleasure, and even languish in the absence of affection. The only difference is that they do it at a much slower rate than we do. So much slower, in fact, that it appears to the ordinary perception that they do not react at all.

"If this is so, then how can we say with certainty that rocks, and even mountain ranges, do not react also as living organisms, but with a reaction time so slow that to catch it with time-lapse photography would require millennia between exposures! Of course, there is no way to prove this, but there is no way of disproving it either. The distinction between living and nonliving is not so easy to make."

"That's clever," we think, "but from a practical point of view, it cannot be observed that inert matter responds to stimulae, and there is no question that humans do."

"Wrong again!" says de Wit, reading our thoughts. "Any chemist can verify that most chemicals (which usually come out of the ground as rocks) *do* react to stimulation. Under the right conditions, for example, sodium reacts to chlorine (by forming sodium chloride—salt), iron reacts to oxygen (by forming iron oxides—rust), and so on, just as humans react to food when they are hungry and to affection when they are lonely."

"Well, this is so," we admit, "but it hardly seems fair to compare a chemical reaction to a human reaction. A chemical reaction either happens or it does not happen. There is nothing in between. When two such chemicals are combined properly, they react; if they are not properly combined, they do not react. Humans are much more complex.

"If we offer food to a hungry person, he might eat it or he might not, depending upon his circumstances; and if he eats, he might eat his fill or he might not. Consider the person who is hungry and late for an appointment. If the appointment is important enough, he will go without eating, even though he is hungry. If a person knows that his food is poisonous, he will not eat, even though he is hungry. It is a matter of processing information and responding appropriately that distinguishes a human reaction from a chemical reaction. Chemicals have no options; they always must act one way or the other."

"Of course," beams Jim de Wit, "but how do we know that our responses are not as rigidly preprogrammed as those of a chemical, with the only difference being that our programs are enormously more complex? We may not have any more freedom of action than stones do, although, unlike stones, we deceive ourselves into thinking that we do!"

We have no way to dispute this argument. De Wit has shown us the arbitrary quality of our prejudices. We would like to think that we are different from stones because we are living and they are not, but there is no way we can prove our position or disprove his. We cannot establish clearly that we are different from inorganic substances. That means that, logically, we must admit that we may not be alive. Since this is absurd, the only alternative is to admit that "inanimate" objects may be living.

The distinction between organic and inorganic is a conceptual prejudice. It becomes even harder to maintain as we advance into quantum mechanics. Something is organic, according to our definition, if it can respond to processed information. The astounding discovery awaiting newcomers to physics is that the evidence gathered in the development of quantum mechanics indicates that subatomic "particles" constantly appear to be making decisions! More than that, the decisions they seem to make are based on decisions made elsewhere. Subatomic particles seem to know *instantaneously* what decisions are made elsewhere, and elsewhere can be as far away as another galaxy! The key word is *instantaneously*. How can a subatomic particle over here know what decision another particle over there has made *at the same time the particle over there makes it*? All the evidence belies the fact that quantum particles are actually particles.

A particle, as we mentally picture it (classically defined) is a thing which is confined to a region in space. It is not spread out. It is either here or it is there, but it cannot be both here *and* there at the same time.

A particle over here can communicate with a particle over there (by shouting at it, sending it a TV picture, waving, etc.), but that takes *time,* even if only milliseconds. If the two particles are in different galaxies, it could take centuries. For a particle here to know what is going on over there while it is happening, it must *be* over there. But if it is over there, it cannot be here. If it is both places at once, then it is no longer a particle.

This means that "particles" may not be particles at all (page 34). It also means that these apparent particles are related with other particles in a dynamic and intimate way that coincides with our definition of organic.

Some biologists believe that a single plant cell carries within it the capability to reproduce the entire plant. Similarly, the philosophical implication of quantum mechanics is that all of the things in our universe (including us) that appear to exist independently are actually parts of one all-encompassing organic pattern, and that no parts of that pattern are ever really separate from it or from each other.

To understand these decisions and what makes them, let us start with a discovery made in 1900 by Max Planck. This year generally is considered the birthday of quantum mechanics. In December of that year, Planck reluctantly presented to the scientific community a paper which was to make him famous. He himself was displeased with the implications of his paper, and he hoped that his colleagues could do what he could not do: explain its contents in terms of Newtonian physics. He knew in his heart, however, that they could not, and that neither could anyone else. He also sensed, and correctly so, that his paper would shift the very foundations of science.

What had Planck discovered that disturbed him so much? Planck had discovered that the basic structure of nature is granular, or, as physicists like to say, discontinuous.

What is meant by "discontinuous"?

If we talk about the population of a city, it is evident that it can fluctuate only by a whole number of people. The least the population of a city can increase or decrease is by one person. It cannot increase by .7 of a person. It can increase or decrease by fifteen people, but not by 15.27 people. In the dialect of physics, a population can change only in discrete increments, or discontinuously. It can get larger or smaller only in jumps, and the smallest jump that it can make is a whole person. In general, this is what Planck discovered about the processes of nature.

Planck did not intend to undermine the foundations of Newtonian physics. He was a conservative German physicist. Rather, he inadvertently fathered the revolution of quantum mechanics by attempting to solve a specific problem dealing with energy radiation.

Planck was searching for an explanation of why things behave as they do when they get hot. Namely, he wanted to know how objects glow brighter as they get hotter, and change color when the temperature is increased or decreased.

Classical physics, which successfully had unified such diverse fields as acoustics, optics, and astronomy, which had all but satiated the scientific appetite, which had unraveled the enigmas of the universe and rearranged them in neat packages, held no sensible explanation of this commonplace phenomenon. It was, to use the parlance of the day, one of the few "clouds" on the horizon of classical physics.

In 1900 physicists pictured the atom as a nucleus that looked something like a plum to which were attached tiny-protruding springs. (This was before the planetary model of the atom.) At the end of each spring was an electron. Giving the atom a jolt, by heating it, for instance, caused its electrons to jiggle (oscillate) on the ends of their springs. The jiggling electrons were thought to give off radiant energy, and this was thought to account for the fact that hot objects glow. (An accelerating electrical charge creates electromagnetic radiation.) (An electron carries an electrical charge [negative] and if it is jiggling, it is accelerating—first in one direction and then in the other.)

Physicists thought that heating the atoms in a metal caused them to become agitated, and this in turn caused their electrons to jiggle up and down and emit light in the process. The energy that the atom absorbed when it was jolted (heated), the theory went, was radiated by the jiggling electrons. (You can substitute "atomic oscillators" if your friends won't take "jiggling electrons" seriously.)

This same theory also claimed that the energy absorbed by an atom was distributed equally to its oscillators (electrons) and that those electrons which oscillated (jiggled) at higher frequencies (faster) radiated their energy most efficiently.

Unfortunately, this theory didn't work. It "proved" some very incorrect things. First, it "proved" that all heated objects emit more high-frequency light (blue, violet) than low-frequency light (red). In other words, even moderately hot objects, according to this classical theory, emit an intense blue-white color, just like objects which are white-hot, but in lesser amounts. This is incorrect. Moderately hot objects emit primarily red light. Second, the classical theory "proved" that highly heated objects radiate infinite amounts of high-frequency light. This is incorrect. Highly heated objects emit a finite amount of high-frequency light.

Do not be concerned with high frequencies and low frequencies. These terms will be explained shortly. The point is that Planck was exploring one of the last major problems of classical physics: its erroneous predictions concerning energy radiation. Physicists dubbed this problem "The Ultra-Violet Catastrophe." Although it sounds like a rock band, "The Ultra-Violet Catastrophe" reflected a real concern with the fact that heated objects do not radiate large amounts of energy in the form of ultraviolet light (the highest-frequency light known in 1900) the way the classical theory predicted.

The name of the phenomenon that Planck was studying is black-body radiation. Black-body radiation is the radiation that comes from a nonreflecting, perfectly absorbing, flat (nonglossy) black body. Since black is the absence of color (no light is reflected or emitted), black bodies have no color unless we heat them. If a black body is glowing a certain color, we know that it is because of the energy that we have added to it and not because it reflects or emits that color spontaneously.

A "black body" does not always mean a solid body that is black. Suppose that we have a metal box that is completely sealed except for a small hole. If we look inside, what do we see? Nothing, because there is no light in there. (A little light may come in through the hole, but not that much.)

Now suppose that we heat the box until it glows red and then look through the hole. What do we see? Red. (Who said physics is hard?) This is the kind of phenomenon that Planck studied.

All the physicists in 1900 assumed that after the electrons of an excited atom began to jiggle, they radiated their energy smoothly and continuously until they "ran down" and their energy was dissipated. Planck discovered that excited atomic oscillators do not do this. They emit and absorb energy only in specific amounts! Instead of radiating energy smoothly and continuously like a clock spring runs down, they radiate their energy in spurts, dropping to a lower energy level after each spurt until they stop oscillating altogether. In short, Planck discovered that the changes of nature are "explosive," not continuous and smooth.*

Planck was the first physicist to talk about "energy packets" and "quantized oscillators." He sensed that he had made a major discovery, one which ranked with the discoveries of Newton, and he was right. The philosophy and paradigms of physics never were to be the same, although it took another twenty-seven years for "quantum mechanics" to take form.

It is difficult today to understand how bold was Planck's theory of quanta. Victor Guillemin, professor of physics at Harvard, put it this way:

> [Planck] had to make a radical and seemingly absurd assumption, for according to classical laws, and common sense as well,

*". . . the hypothesis of quanta has led to the idea that there are changes in Nature which do not occur continuously but in an explosive manner."—Max Planck, "Neue Bahnen der physikalischen Erkenntnis," 1913, trans., F. d'Albe, *Phil. Mag.* vol. 28, 1914.

it had been presumed that an electronic oscillator, once set in motion by a jolt, radiates its energy smoothly and gradually while its oscillatory motion subsides to rest. Planck had to assume that the oscillator ejects its radiation in sudden spurts, dropping to lesser amplitudes of oscillation with each spurt. He had to postulate that the energy of motion of each oscillator can neither build up nor subside smoothly and gradually but may change only in sudden jumps. In a situation where energy is being transferred to and fro between the oscillators and the light waves, the oscillators must not only *emit* but also *absorb* radiant energy in discrete "packets". . . . He coined the name "quanta" for the packets of energy, and he spoke of the oscillators as being "quantized." Thus, the trenchant concept of the *quantum* entered physical science.[1]

Planck is not only the father of quantum mechanics, he also is the discoverer of Planck's constant. Planck's constant is a certain number which never changes.* It is used to calculate the size of the energy packets (quanta) of each light frequency (color). (The energy in each light quantum of a particular color is the frequency of the light multiplied by Planck's constant.)

All of the energy packets of each color have the same amount of energy. All of the energy packets of red light, for example, are the same size. All of the energy packets of green light are the same size. All of the energy packets of violet light are the same size. The energy packets of violet light, however, are larger than the energy packets of green light, and the energy packets of green light are larger than the energy packets of red light.

In other words, Planck discovered that energy is absorbed and emitted in little chunks and that the size of the chunks of a low-frequency light, like red, is smaller than the size of the chunks of a high-frequency light, like violet. This explains why hot objects radiate energy as they do.

*$h = 6.63 \times 10^{27}$ erg-sec

When a black body is put over low heat, the first color it glows is red because the energy packets of red light are the smallest energy packets in the visible light spectrum. As the heat is increased, more energy is available to shake loose bigger energy packets. The bigger energy packets make the higher-frequency colors, such as blue and violet.

Why does the glow of hot metal seem to increase steadily in brightness as the temperature increases? Because the tiny "steps" upward and downward in brightness are so incredibly small that our eyes cannot discern them. Therefore, on the large scale, or macroscopic level, this aspect of nature is not evident. In the subatomic realm, however, it is the dominant characteristic of nature.

If this discussion of emission and absorption of energy packets reminds you of Niels Bohr (page 14), you are right. However, Bohr was not to arrive at his theory of specific electron orbits for another thirteen years. By that time physicists had discarded the plum-with-jiggling-electrons model of the atom in favor of the planetary model, in which electrons revolve around a nucleus.*

Between Planck's discovery of the quantum (1900) and Bohr's analysis of the hydrogen spectrum (1913), a brilliant physicist burst upon the scene with a force seldom exerted by an individual. His name was Albert Einstein. In one year (1905), at twenty-six, Einstein published five significant papers. Three of them were pivotal in the devel-

---

*Bohr speculated that electronic orbits are arranged by nature at unvarying specific distances from the nucleus of the atom and that, when they absorb energy, the electrons in the atom jump outward from the orbit closest to the nucleus (the "ground state" of the atom) and eventually return to the innermost orbit, in the process emitting energy packets equal to the energy packets that they absorbed in jumping outward. Bohr proposed that when only a little energy is available (low heat), only small energy packets are absorbed by the electrons, and they do not jump out very far. When they return to their lowest energy level, they emit small energy packets, like those of red light. When more energy is available (high heat), larger energy packets are available, the electrons make bigger jumps outward and, on returning, they emit larger energy packets, like those of blue and violet light. Therefore, over low heat, metal glows red, and over high heat, it glows blue-white.

opment of physics, and, to a large extent, in the development of the West. The first of these three papers described the quantum nature of light. It won him a Nobel Prize in 1921. The second paper described molecular motion. The third paper set forth the special theory of relativity, which we will study later.*

Einstein's theory of light was that it is composed of tiny particles. A beam of light, said Einstein, is analogous to a stream of bullets. Each bullet is called a photon. This is similar to what Planck proposed, but actually it is a leap beyond. Planck discovered that energy is absorbed and emitted in packets. He described the *processes* of energy absorption and emission. Einstein theorized that *energy itself* is quantized.

To prove his theory, Einstein referred to a phenomenon called the photoelectric effect. When light hits (impinges on) the surface of a metal, it jars electrons loose from the atoms in the metal and sends them flying off. With appropriate equipment, we can count these electrons and measure how fast they are traveling.

Einstein's theory of the photoelectric effect was that each time one of the bullets, or photons, hits an electron, it knocks it away just as one billiard ball hitting another billiard ball knocks it away.

Einstein based his revolutionary theory on the experimental work of Philippe Lenard (who won the Nobel Prize in 1905). Lenard showed that the flow of electrons in the photoelectric effect begins immediately when the impinging light strikes the target metal. Turn on the light and out come the electrons. According to the wave theory of light, the electrons in a metal only start to jiggle when they are struck by light waves. They do not come out of the metal until they are moving fast enough. This takes several oscillations, like pumping a child's swing higher and higher until it goes around the bar. In short, the wave theory of light predicts a delayed emission of electrons. Lenard's experiments showed a prompt emission of electrons.

---

*Each of Einstein's major 1905 papers dealt with a fundamental physical constant: h, Planck's constant (the photon hypothesis); k, Boltzmann's constant (the analysis of Brownian movement); and c, the velocity of light (the special theory of relativity).

This prompt emission of electrons in the photoelectric effect is explained by Einstein's particle theory of light. Every time a particle of light, a photon, strikes an electron, it immediately knocks it out of its atom.

Lenard also discovered that reducing the intensity of the impinging light beam (making it dimmer) did not reduce the velocity of the rebounding electrons, but it did reduce the number of the rebounding electrons. He found that the velocity of the rebounding electrons could be altered, however, by changing the *color* of the impinging light.

This also was explained by Einstein's new theory. According to Einstein's theory, each photon of a given color, like green, for instance, has a certain amount of energy. Reducing the intensity of a beam of green light only reduces the number of photons in the beam. Each remaining photon, however, still has the same amount of energy as any other photon of green light. Therefore, when any photon of green light strikes an electron, it knocks it away with a certain amount of energy which is characteristic of green-light photons.

Max Planck described Einstein's theory this way:

> . . . The photons (the "drops" of energy) do not grow smaller as the energy of the ray grows less; what happens is that their magnitude remains unchanged and they follow each other at greater intervals.[2]

Einstein's theory also substantiated Planck's revolutionary discovery. High-frequency light, like violet, is made of higher-energy photons than low-frequency light, like red. Therefore, when violet light, which is made of high-energy photons, strikes an electron, it causes the electron to rebound with a high velocity. When red light, which is made of low-energy photons, strikes an electron, it causes the electron to rebound at a low velocity. In either case, increasing or decreasing the intensity of the light increases or decreases the number of rebounding electrons, but only by changing the color of the impinging light can we change their velocity.

In short, Einstein demonstrated, using the photoelectric effect, that light is made of particles, or photons, and that the photons of high-frequency light have more energy than the photons of low-frequency light. This was a momentous achievement. The only problem was that one hundred and two years earlier an Englishman named Thomas Young had shown that light is made of waves, and no one, including Einstein, was able to disprove him.

Now we come to the matter (no pun) of waves. A particle is something that is contained in one place. A wave is something that is spread out. Here are some types of waves.

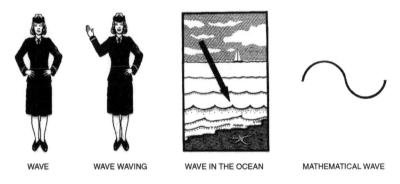

WAVE   WAVE WAVING   WAVE IN THE OCEAN   MATHEMATICAL WAVE

We are concerned only with the last type of wave. Here is a more detailed picture of it.

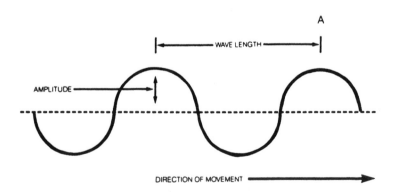

A

WAVE LENGTH

AMPLITUDE

DIRECTION OF MOVEMENT

A *wavelength* is the distance between one crest of a wave and the next. The longest radio waves are over six miles long. X-rays, on the other hand, are only about one billionth of a centimeter long. Visible light has wavelengths in the neighborhood of four to eight one hundred thousandths of a centimeter.

The *amplitude* of a wave is the height of the wave crest above the dotted line. Here are three waves with different amplitudes. The one in the middle has the largest amplitude.

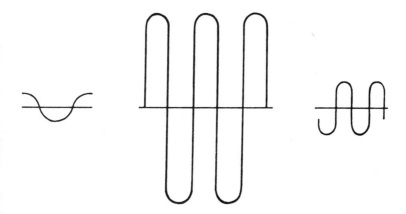

The *frequency* of the wave tells us how many crests pass a given point (like Point A in the drawing) each second. If the wave is moving in the direction of the arrow and a crest passes point A each second, the frequency of the wave is one cycle per second. If ten and one half crests pass point A every second, the frequency of the wave is 10.5 cycles per second. If ten thousand crests pass the same point every second, the frequency of the wave is 10,000 cycles per second, and so on.

The velocity of the wave can be determined by multiplying the wavelength by the frequency. For example, if the wavelength of a wave is two feet and the frequency of the wave is one cycle per second, the wave is moving one wavelength (two feet) every second. Therefore, its velocity is two feet per second. If the wavelength is two feet and the frequency is three cycles per second, the velocity of the wave is six feet per second because the wave moves three wavelengths forward every second.

There is nothing complicated about this. We can determine how fast a man is running if we know the length of his stride and how many of them he takes in a second. By multiplying them together we get how far the man runs in a second. If his stride is three feet and he takes two strides per second, then he runs six feet per second (about four miles per hour). We do the same things with waves, except that we use wavelengths instead of strides.

Although the velocity of a light wave *can* be determined by multiplying its wavelength by its frequency, it is not necessary. Physicists have discovered that the velocity of light in empty space is *always* 186,000 miles per second. This applies to all electromagnetic waves, including light. Therefore, all light waves (blue ones, green ones, red ones, etc.) have the same velocity as radio waves, x-rays, and all the other forms of electromagnetic radiation. The speed of light is a constant. It is represented by the letter "c."

The constant "c" is (approximately) 186,000 miles per second and it never varies (which is what makes it a "constant"). It does not matter whether light is going up or down, has a high frequency or a low frequency, a large wavelength or a small wavelength, is coming toward us or going away from us: Its velocity is always 186,000 miles per second. This fact led Albert Einstein to the theory of special relativity, as we shall see later.

It also permits us to know both the frequency and the wavelength of light if we know either one of them. This is because the product of the two is always 186,000 miles per second in empty space. The larger one of them is, the smaller the other must be. For example, if we know that by multiplying two numbers together we get 12 for an answer, and if we know that one of the numbers is 6, then we also know that the other number *must* be 2. If we know that one of the numbers is 3, then we know that the other number *must* be 4.

Similarly, the higher the frequency of a light wave, the shorter its wavelength must be; the lower the frequency of a light wave, the

longer its wavelength must be. In other words, high-frequency light
has a short wavelength and low-frequency light has a long wavelength.

Now we return to Planck's discovery. Planck discovered that the
energy of a light quantum increases with frequency. The higher the
frequency, the higher the energy. Energy is proportional to frequency,
and Planck's constant is the "constant of proportionality" between
them. This simple relation between frequency and energy is important. It is central to quantum mechanics. The higher the frequency,
the higher the energy; the lower the frequency, the lower the energy.

When we put wave mechanics and Planck's discovery together
we get this: High-frequency light, such as violet light, has a short
wavelength and high energy; low-frequency light, such as red light,
has a long wavelength and low energy.

This explains the photoelectric effect. Photons of violet light
knock electrons loose from the atoms of a metal and send them flying
away at a higher velocity than photons of red light because the photons of violet light, which is a high-frequency light, have more energy
than the photons of red light, which is a low-frequency light.

This all makes sense if you overlook the fact that we are talking
about particles (photons) in terms of waves (frequencies) and waves
in terms of particles, which, of course, makes no sense at all.

If you feel that you understand the last few pages, congratula-
tions! You have mastered the most difficult mathematics in the book.
If not, go back to page 60 and reread these pages. It is easy to dance
with wavelengths and frequencies if you know how they are connected.

Waves are playful creatures that like to do dances of their own. For
example, under certain conditions they bend around corners. When
this happens it is called diffraction.

Imagine that we are in a helicopter hovering over the mouth of
an artificial harbor. The mouth of the harbor is wide enough for two
aircraft carriers to pass each other going through it. The sea is rough
and the wind and waves are blowing straight into the mouth of the

harbor. When we look down, this is the pattern that we see the waves making in the harbor:

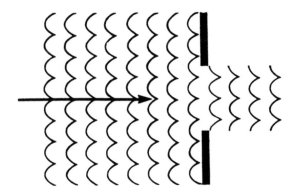

The waves are stopped cleanly by the walls of the harbor except at the harbor entrance, where they continue straight forward into the harbor until they are dissipated.

Now imagine that the mouth of the harbor is so small that a rowboat scarcely can pass through it. As we look down from the helicopter, the pattern we see is quite different.

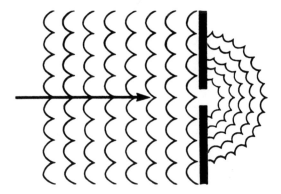

Instead of moving directly ahead into the harbor, the waves inside the harbor are spreading out from the mouth of the harbor almost as if it were a pond and we had dropped a rock into it at that point. This is diffraction.

Why does it happen? Why does reducing the size of the harbor mouth cause the wave pattern inside the harbor to spread out in semi-circles?

The answer lies in comparing the size of the harbor mouth to the size of the wavelength of the incoming waves. In the first case, the size of the harbor mouth is considerably larger than the distance between the crests of the waves going through it, and the waves proceed directly through it into the harbor following a straight line (recti-linear propagation) as waves usually do.

In the second case, the size of the harbor mouth is about the same size, or smaller, than the wavelength of the incoming waves, and when this happens, it causes the characteristic pattern (diffraction) that we see in the drawing.

Whenever waves pass through an opening that is so small that the wavelength of the waves passing through it seems large in comparison, the waves passing through it diffract.

Since light is a wave phenomenon (according to the wave theory of light), it should behave in the same way, and it does. If we place a light source behind a cut-out like the one below, it casts a projection on the wall like this one:

This is analogous to sea waves entering the large harbor mouth. The width of the cut-out is millions of times larger than the wavelength of the light. As a result, the light waves go straight through it, following straight lines and projecting onto the wall a figure with the same shape as the cut-out. Notice especially that this projection has distinct borders between the bright area and the dark area.

If we make the cut-out no larger than a razor slit so that its width is roughly as small as the wavelength of the incoming light, the light diffracts. Now the sharp boundary between the light area and the dark area disappears and we see a bright area that fades into darkness at the edges. Instead of proceeding in a straight line to the wall, the light beam has spread out like a fan. This is diffracted light.

Now that you know the punch line, here comes the story.

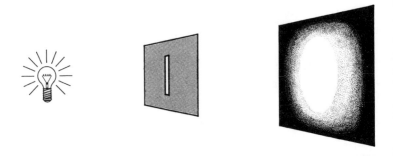

In 1803, Thomas Young settled once and for all (he thought) the question of the nature of light. He used an experiment that was both simple and dramatic. In front of a light source (Young used sunlight coming through a hole in a screen) he placed a screen with two vertical slits in it. Each slit could be covered over with a piece of material.

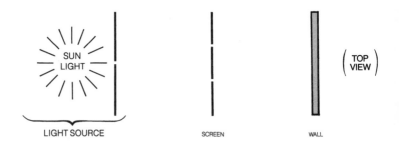

LIGHT SOURCE   SCREEN   WALL

On the other side of the double-slit screen was a wall against which the light coming through the double slits could shine. When the light source was turned on and one of the slits was covered up, the wall was illuminated like the drawing below.

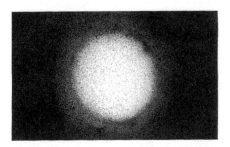

When both slits were uncovered, however, Young made history. The projection on the wall should have been the sum of the light from the two slits, but it wasn't. Instead, the wall was illuminated with alternating bands of light and darkness! The center band was the brightest. On both sides of the center band of light were bands of darkness; then bands of light, but less intense than the center band; then bands of darkness, etc., as below.

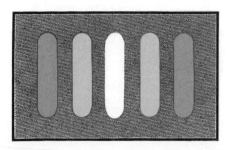

How could this happen?

The simplicity of the answer is what makes this experiment a great one. The alternating light and dark bands are a well-known phenomenon of wave mechanics called interference. Interference results when the waves of light diffracting from the two slits interfere with each other. In some places these waves overlap and reinforce one another. In other places they cancel each other.

In areas where one wave crest overlaps another wave crest, the result is an intensification of light (the light bands). In areas where a crest meets a trough, they cancel each other and no light reaches the wall (the dark bands).

It is just as if we dropped two stones into a pond simultaneously and watched the waves spreading from their points of entry. The waves that the stones make interfere with each other. In places where the crests of the waves caused by one stone meet the crests of the waves caused by the other stone, large waves result. In places where the troughs of the waves caused by one stone meet the crests of the waves caused by the other stone, the water is calm.

In short, Young's double-slit experiment showed that light must be wave-like because only waves can create interference patterns. The situation, then, was as follows: Einstein, using the photoelectric effect, "proved" that light is particle-like and Young, using the phenomenon of interference, "proved" that light is wave-like. But a wave cannot be a particle and a particle cannot be a wave.

That is just the beginning! Since Einstein "proved" that light is composed of photons, let us go back to Young's double-slit experiment and run it with photons.* (This has been done.) Suppose that we have a light gun which can fire, in effect, one photon at a time. The experiment is set up as before, except that only one slit is open. Now we fire the photon, it goes through the open slit, and we mark where it hits the wall (using a photographic plate). Because we have done this experiment before, we notice that the photon has landed in an area that would be dark if the second slit were open. That is, if the second slit were open, no photons would be recorded in this area.

To make sure, we do the experiment again, but this time we leave both of the slits open. Just as we thought, there are no photons recorded now in the area where the photon hit in our first experiment.

*If we assume a particle aspect in the double-slit experiment we will violate the uncertainty relation unless we also assume nonlocality.

When both slits are open and interference is present, this area is in the middle of a dark band.

The question is, *How did the photon in the first experiment know that the second slit was not open?* Think about it. If both slits are open, there are *always* alternating bands of illuminated and dark areas. This means that there are always areas where the photons never go (otherwise there would not be any dark areas). If one of the slits is closed, there is no interference and the dark bands disappear; the whole wall becomes illuminated, including those areas which previously were dark when both slits were open.

When we fired our photon and it went through the first slit, how did it "know" that it could go to an area that must be dark if the other slit were open? In other words, how did the photon know that the other slit was closed?

> "The central mystery of quantum theory," wrote Henry Stapp, is 'How does information get around so quick?' " How does the particle know that there are two slits? How does the information about what is happening everywhere else get collected to determine what is likely to happen here?[3]

There is no definitive answer to this question. Some physicists, like E. H. Walker, speculate that photons may be *conscious!*

> Consciousness may be associated with all quantum mechanical processes . . . since everything that occurs is ultimately the result of one or more quantum mechanical events, the universe is "inhabited" by an almost unlimited number of rather discrete conscious, usually nonthinking entities that are responsible for the detailed working of the universe.[4]

Whether Walker is correct or not, it appears that if there really are photons (and the photoelectric effect "proves" that there are), then it also appears that the photons in the double-slit experiment somehow "know" whether or not both slits are open and that they act accordingly.*

*An explanation other than "knowing" might be synchronicity, Jung's acausal connecting principle.

This brings us back to where we started: Something is "organic" if it has the ability to process information and to act accordingly. We have little choice but to acknowledge that photons, which are energy, do appear to process information and to act accordingly, and that therefore, strange as it may sound, they seem to be organic. Since we are also organic, there is a possibility that by studying photons (and other energy quanta) we may learn something about us.

The wave-particle duality was the end of the line for classical causality. According to that way of thinking, if we know certain initial conditions, we can predict the future of events because we know the laws that govern them. In double-slit experiments we know all that we can know about initial conditions and we still can't predict correctly what happens to single photons.

In experiment one, for example (only one slit open), we know the origin of the photon (the lamp), its velocity (186,000 miles per second), and its direction just prior to passing through the open slit. Using Newton's laws of motion, we can predict where the photon will land on the photographic plate. Let us suppose that we make these calculations.

Now let us consider experiment two (both slits open). Again we know the origin of the photon, it velocity, and its direction just prior to passing through the open slit. The initial conditions of the photon in experiment one are the same as those of the photon in experiment two. They both start from the same place, travel at the same speed, go to the same place, and therefore, are moving in the same direction just prior to passing through slit number one. The only difference is that in the second experiment, the second slit also is open. Again, using Newton's laws of motion, let us calculate where the photon will land on the photographic plate.

Since we used the same figures and the same formulas in both cases, we get identical answers indicating that the photon in experiment one will impact in exactly the same place as the photon in experiment two. That is the problem. The photon in experiment two will *not* impact in the same area as the photon in experiment one because the photon in experiment one landed in an area that is a dark band in experiment two. In other words, the two photons do not go to the same place even though the initial conditions pertaining to both of them are identical and known to us.

We cannot determine the paths of individual photons. We can determine what the wave pattern on the wall will be, but in this case we are interested in a single photon, not waves of them. In other words, we know the pattern that large groups of photons will make, and their distribution in the pattern, but we have no way of knowing which photons will go where. All that we can say about a single photon is the probability of finding it in a given place.

The wave-particle duality was (is) one of the thorniest problems in quantum mechanics. Physicists like to have tidy theories which explain everything, and if they are not able to do that, they like to have tidy theories about why they can't. The wave-particle duality is not a tidy situation. In fact, its untidiness has forced physicists into radical new ways of perceiving physical reality. These new perceptual frames are considerably more compatible with the nature of personal experience than were the old.

For most of us, life is seldom black and white. The wave-particle duality marked the end of the "Either-Or" way of looking at the world. Physicists no longer could accept the proposition that light is *either* a particle *or* a wave because they had "proved" to themselves that it was *both*, depending on how they looked at it.

Of course, Einstein was aware of the fact that his photon theory contradicted Young's wave theory without disproving it. He speculated that photons were guided by "ghost waves." Ghost waves were mathematical entities which had no actual existence. The photons seemed to follow paths which had all the mathematical characteristics of waves, but which in reality did not exist. Some physicists still view the wave-particle paradox this way, but for most physicists, this explanation seems too contrived. It is an answer which appears to make sense, but somehow doesn't explain anything.

The wave-particle duality prompted the first real step in understanding the newly unfolding quantum theory. In 1924, Bohr and two of his colleagues, H. A. Kramers and John Slater, suggested that the waves in question were *probability waves*. Probability waves were mathematical entities by which physicists could predict the probability of certain events occurring or not occurring. Their mathematics did not prove correct, but their idea, which was unlike anything that had been proposed before, was sound. Later, with a different formalism (mathematical structure), the idea of probability waves developed into one of the distinguishing characteristics of quantum mechanics.

Probability waves, as Bohr, Kramers, and Slater thought of them, was an entirely new idea. Probability itself was not new, but this type of probability was. It referred to what somehow already was happening, but had not yet been actualized. It referred to a *tendency* to happen, a tendency that in an undefined way existed of itself, even if it never became an event. Probability waves were mathematical catalogues of these tendencies.

This was something quite different from classical probability. If we throw a die in a casino, we know, using classical probability, that the chances of getting the number that we want is one in six. The probability wave of Bohr, Kramers, and Slater meant much more than that.

According to Heisenberg:

> It meant a tendency for something. It was a quantitative version of the old concept of "potentia" in Aristotelian philosophy. It introduced something standing in the middle between the idea of an event and the actual event, a strange kind of physical reality just in the middle between possibility and reality.[5]

By 1924, Planck's discovery of the quantum was producing seismic effects in physics. It enabled Einstein to discover the photon, which caused the wave-particle duality, which led to probability waves. The physics of Newton was a thing of the past.

Physicists found themselves dealing with energy that somehow processed information (which made it organic), and unaccountably presented itself in patterns (waves). In short, physicists found themselves dealing with Wu Li—patterns of organic energy.

1

What Happens

Quantum mechanics is a procedure. It is a specific way of looking at a specific part of reality. The only people who use it are physicists. The advantage of following the procedure of quantum mechanics is that it allows us to predict the *probabilities* of certain results provided our experiment is performed in a certain way. The purpose of quantum mechanics is not to predict what actually will happen, but only to predict the probabilities of various possible results. Physicists would like to be able to predict subatomic events more accurately, but, at present, quantum mechanics is the only workable theory of subatomic phenomena that they have been able to construct.

Probabilities follow deterministic laws in the same way that macroscopic events follow deterministic laws. There is a direct parallel. If we know enough about the initial conditions of an experiment, we can calculate, using rigid laws of development, exactly what the probability is for a certain result to occur.

For example, there is no way that we can calculate where a single photon in a double-slit experiment will strike the photographic plate (page 70–71). However, we can calculate with precision the probability that it will strike it at a certain place, provided that the experiment has been prepared properly and that the results are measured properly.

Suppose that we calculate a 60 percent probability for the photon to land in area A. Does that mean that it can land somewhere else? Yes. In fact, there is a 40 percent probability that it will.

In that case (asking the question for Jim de Wit), what determines where the photon will land? The answer given by quantum theory: pure chance.

This pure-chance aspect was another objection that Einstein had about quantum mechanics. It is one of the reasons that he never accepted it as the fundamental physical theory. "Quantum mechanics is very impressive," he wrote in a letter to Max Born, ". . . but I am convinced that God does not play dice."[1]

Two generations later, J. S. Bell, a Scotish physicist, proved that he may have been right, but that is another story, which we will come to later.

The first step in the procedure of quantum mechanics is to prepare a physical system (the experimental apparatus) according to certain specifications, in an area called the region of preparation.

The second step in the procedure of quantum mechanics is to prepare another physical system to measure the results of the experiment. This measuring system is located in an area called the region of measurement. Ideally, the region of measurement is far away from the region of preparation. Of course, to a subatomic particle, even a small macroscopic distance is a long way.

Now let us perform the double-slit experiment using this procedure. First, we set a light source on a table, and then, a short distance away, we place a screen with two vertical slits in it. The area where these apparatuses are located is the region of preparation. Next we fix an unexposed photographic plate on the opposite side of the screen from the light source. This area is the region of measurement.

The third step in the procedure of quantum mechanics is to translate what we know about the apparatus in the region of preparation (the light and the screen) into mathematical terms which represent it, and to do likewise for the apparatus that is located in the region of measurement (the photographic plate).

To do this we need to know the specifications of the apparatus.

In practice, this means that we give the technician who sets up the equipment precise instructions. We tell him, for example, the exact distance to place the double-slit screen from the light source, the frequency and intensity of the light that we will use, the dimensions of the two slits and their position relative to each other and to the light source, etc. We also give him equally explicit instructions concerning the measuring apparatus, such as where to put it, the type of photographic film that we will use, how to develop it, etc.

After we translate these specifications of the experimental arrangement into the mathematical language of quantum theory, we feed these mathematical quantities into an equation that expresses the form of natural causal development. Notice that this last sentence doesn't say anything about what is developing. That is because nobody knows. The Copenhagen Interpretation of Quantum Mechanics (page 40) says that quantum theory is a complete theory because it works (correlates experience) in every possible experimental situation, not because it explains in detail what is going on.* (Einstein's complaint was that quantum theory doesn't fully explain things because it deals with group behavior and not with individual events.)

However, when it comes to predicting group behavior, quantum theory works as advertised. In a double-slit experiment, for example, it can predict exactly the probabilities of a photon being recorded in region A, in region B, in region C, and so forth.

Of course, the last step in the procedure of quantum mechanics is actually to do the experiment and get a result.

To apply quantum theory, the physical world must be divided into two parts. These parts are the observed system and the observing system. The observed system and the observing system are not the same as the region of preparation and the region of measurement. "Region of preparation" and "region of measurement" are terms

---

*According to the complementarity argument, which is at the heart of the Copenhagen Interpretation, the latitude in the choice of possible wave functions exactly corresponds to (or at least includes) the latitude in the set of possible experimental arrangements, so that every possible experimental situation or arrangement is covered by quantum theory.

which describe the physical organization of the experimental apparatus. "Observed system" and "observing system" are terms which pertain to the way that physicists analyze the experiment. (The "observed" system, by the way, cannot be observed until it interacts with the observing system, and even then all that we can observe are its effects on a measuring device.)

The observed system in the double-slit experiment is a photon. It is pictured as traveling between the region of preparation and the region of measurement. The observing system in all quantum mechanical experiments is the environment which surrounds the observed system—including the physicists who are studying the experiment. While the observed system is traveling undisturbed ("propagating in isolation"), it develops according to a natural causal law. This law of causal development is called the Schrödinger wave equation. The information that we put into the Schrödinger wave equation is the data about the experimental apparatuses that we have transcribed into the mathematical language of quantum theory.

Each set of these experimental specifications that we transcribed into the mathematical language of quantum theory corresponds to what physicists call an "observable." Observables are the features of the experiment and nature that are considered to be fixed, or determined, when and if the experimental specifications that we have transcribed actually are met. We may have transcribed into mathematical language several experimental specifications for the region of measurement, each one corresponding to a different possible result (the possibility that the photon will land in region A, the possibility that the photon will land in region B, the possibility that the photon will land in region C, etc.).

In the world of mathematics, the experimental specifications of each of these possible situations in the region of measurement and in the region of preparation corresponds to an observable.* In the world

---

*Each set of experimental specifications A or B, that can be transcribed into a corresponding theoretical description $\mathcal{S}$ A or $\mathcal{S}$ B, corresponds to an observable. In the mathematical theory the observable is $\mathcal{S}$ A or $\mathcal{S}$ B; in the world of our experience the observable is the possible occurrence (coming into our experience) of the satisfied specifications.

of experience, an observable is the possible occurrence (coming into our experience) of one of these sets of specifications.

In other words, what happens to the observed system between the region of preparation and the region of measurement is expressed mathematically as a *correlation* between two observables (production and detection). Yet we know that the observed system is a particle—a photon. Said another way, the photon is a *relationship* between two observables. This is a long, long way from the building-brick concept of elementary particles. For centuries scientists have tried to reduce reality to indivisible entities. Imagine how surprising and frustrating it is for them to come so close (a photon is very "elementary"), only to discover that elementary particles don't have an existence of their own!

As Stapp wrote for the Atomic Energy Commission:

> . . . an elementary particle is not an independently existing, unanalyzable entity. It is, in essence, a set of relationships that reach outward to other things.[2]

Furthermore, the mathematical picture which physicists have constructed of this "set of relationships" is very similar to the mathematical picture of a real (physical) moving particle.* The motion of such a set of relationships is governed by exactly the same equation which governs the motion of a real moving particle.

Wrote Stapp:

> A long-range correlation between observables has the interesting property that the equation of motion which governs the propagation of this effect is precisely the equation of motion of a freely moving particle.[3]

Things are not "correlated" in nature. In nature, things are as they are. Period. "Correlation" is a concept which *we* use to describe

---

*The particle is represented by a wave function which has *almost* all of the characteristics (when properly squared, to get a probability function) of a probability density function. However, it lacks the crucial feature of a probability density function, namely the property of being positive.

connections which *we* perceive. There is no word, "correlation," apart from people. There is no concept, "correlation," apart from people. This is because only people use words and concepts.

"Correlation" is a concept. Subatomic particles are correlations. If we weren't here to make them, there would not be any concepts, including the concept of "correlation." In short, if we weren't here to make them, there wouldn't be any particles!*

Quantum mechanics is based on the development in isolation of an observed system. "Development in isolation" refers to the isolation that we create by separating the region of preparation from the region of measurement. We call this situation "isolation," but in reality, nothing is completely isolated, except, perhaps, the universe as a whole. (What would it be isolated from?)

The "isolation" that we create is an idealization, and one point of view is that quantum mechanics allows us to idealize a photon from the fundamental unbroken unity so that we can study it. In fact, a "photon" seems to become isolated from the fundamental unbroken unity *because* we are studying it.

Photons do not exist by themselves. All that exists by itself is an unbroken wholeness that presents itself to us as webs (more patterns) of relations. Individual entities are idealizations which are correlations made by us.

In short, the physical world, according to quantum mechanics, is:

*From the pragmatic point of view, nothing can be said about the world "out there" except via our concepts. However, even within the world of our concepts particles do not seem to have an independent existence. They are represented in theory only by wave functions and the meaning of the wave function lies only in correlations of other (macroscopic) things.

Macroscopic objects, like a "table" or a "chair," have certain direct experiential meanings, that is, we organize our sensory experiences directly in terms of them. These experiences are such that we can believe that these objects have a persisting existence and well-defined location in space-time that is logically independent of other things. Nonetheless, the concept of independent existence evaporates when we go down to the level of particles. This limitation of the concept of independent entity at the level of particles emphasizes, according to the pragmatic view, that even tables and chairs are, for us, tools for correlating experience.

> . . . not a structure built out of independently existing unana-
> lyzable entities, but rather a web of relationships between ele-
> ments whose meanings arise wholly from their relationships to
> the whole. (Stapp)[4]

The new physics sounds very much like old eastern mysticism.

What happens between the region of preparation and the region of
measurement is a dynamic (changing with time) unfolding of possibilities that occurs according to the Schrödinger wave equation. We can
determine, for any moment in the development of these possibilities,
the probability of any one of them occurring.

One possibility may be that the photon will land in region A.
Another possibility may be that the photon will land in region B. However, it is not possible for the same photon to land in region A and in
region B at the same time. When one of these possibilities is actualized, the probability that the other one will occur at the same time
becomes zero.

How do we cause a possibility to become an actuality? We "make
a measurement." Making a measurement interferes with the development of these possibilities. In other words, making a measurement
interferes with the development in isolation of the observed system.
When we interfere with the development in isolation of the observed
system (which is what Schrödinger's wave equation governs) we actualize one of the several potentialities that were a part of the observed
system while it was in isolation. For example, as soon as we detect the
photon in region A, the possibility that it is in region B, or anyplace
else, becomes nihil.

The development of possibilities that takes place between the
region of preparation and the region of measurement is represented
by a particular kind of mathematical entity. Physicists call this mathematical entity a "wave function" because it looks, mathematically, like
a development of waves which constantly change and proliferate. In a
nutshell, the Schrödinger wave equation governs the development in

isolation (between the region of preparation and the region of measurement) of the observed system (a photon in this case) which is represented mathematically by a wave function.

A wave function is a mathematical fiction that represents all the possibilities that can happen to an observed system when it interacts with an observing system (a measuring device). The form of the wave function of an observed system can be calculated via the Schrödinger wave equation for any moment between the time the observed system leaves the region of preparation and the time that it interacts with the observing system.

Once the wave function is calculated, we can perform a simple mathematical operation on it (square its amplitude) to create a second mathematical entity called a probability function (or, technically, a "probability density function"). The probability function tells us the probabilities at a given time(s) of each of the possibilities represented by the wave function. The wave function is calculated with the Schrödinger wave equation. It deals with possibilities. The probability function is based upon the wave function. It deals with probabilities.

There is a difference between possible and probable. Some things may be possible, but not very probable, like snow falling in the summer, except in Antarctica where it is both possible and probable.

The wave function of an observed system is a mathematical catalogue which gives a physical description of those things which could happen to the observed system when we make a measurement on it. The probability function gives the probabilities of those events actually happening. It says, "These are the odds that this or that will happen."

Before we interfere with the development in isolation of an observed system, it merrily continues to generate possibilities in accordance with the Schrödinger wave equation. As soon as we make a measurement, however—look to see what is happening—the probability of all the possibilities, except one, becomes *zero,* and the probability of that possibility becomes *one,* which means that it happens.

The development of the wave function (possibilities) follows an unvarying determinism. We calculate this development by using the

Schrödinger wave equation. Since the probability function is based upon the wave function, the probabilities of possible happenings also develop deterministically via the Schrödinger wave equation.

This is why we can predict accurately the probability of an event, but not the event itself. We can calculate the probability of a desired result, but when we make a measurement, that result may or may not be the one that we get. The photon may land in region B or it may land in region A. Which possibility becomes reality is, according to quantum theory, a matter of chance.

Now back to the double-slit experiment. We cannot predict where a photon in a double-slit experiment will land. However, we can calculate where it is most likely to land, where it is next likely to land, and so on.* This is how it happens.

Suppose that we place a photon detector at slit one and another photon detector at slit two. Now we emit photons from the light source. Sooner or later one of them will go through one slit or the other. There are two possibilities for that photon. It can go through slit one and detector one will fire, or it can go through slit two and detector two will fire. Each of these possibilities is included in the wave function of that photon.

Let us say that when we examine the detectors we find that detector two has fired. As soon as we know this we also know that the photon did not go through slit one. That possibility no longer exists, and, therefore, the wave function of the photon has changed.

The graphic representation (picture) of the wave function of the photon, before we made the measurement, had two humps in it. One of the humps represented the possibility of the photon passing through slit one and detector one firing. The other hump represented the possibility of the photon passing through slit two and detector two firing.

---

*What we can predict is the probability corresponding to any specification that can be mapped into a density function. Accurately speaking, we do not calculate probabilities at points, but rather transition probabilities between two states (initial preparation, final detection), each of which is represented by a continuous function of x and p (position and momentum).

When the photon was detected passing through slit two, the possibility that it would go through slit one ceased to exist. When that happened, the hump in the graphic representation of the wave function representing that possibility changed to a straight line. This phenomenon is called the "collapse of the wave function."

Physicists speak as if the wave function exhibits two very different modes of development. The first is a smooth and dynamic development, which we can predict because it follows the Schrödinger wave equation. The second is abrupt and discontinuous (that word, again). This mode of development is the collapse of the wave function. Which part of the wave function collapses is a matter of chance. The transition from the first mode to the second mode is called a quantum jump.

The Quantum Jump is not a dance. It is the abrupt collapse of all the developing aspects of the wave function except the one that actualizes. The mathematical representation of the observed system literally leaps from one situation to another, with no apparent development between the two.

In a quantum mechanical experiment, the observed system, traveling undisturbed between the region of preparation and the region of measurement, develops according to the Schrödinger wave equation. During this time, all of the allowed things that could happen to it unfold as a developing wave function. However, as soon as it interacts with a measuring device (the observing system), one of those possibilities actualizes and the rest cease to exist. The quantum leap is from a multifaceted potentiality to a single actuality.

The quantum leap is also a leap from a reality with a theoretically infinite number of dimensions into a reality which has only three. This is because the wave function of the observed system, before it is observed, proliferates in many mathematical dimensions.

Take the wave function of our photon in the double-slit experiment for example. It contains two possibilities. The first possibility is that the photon will go through slit one and detector one will fire, and the second possibility is that the photon will go through slit two and detector two will fire. Each of these possibilities, alone, would be represented by a wave function that exists in three dimensions and a time. This is because our reality has three dimensions, length, width, and depth, along with time.

If we want to describe a physical event accurately, we must say where it happened and when.

To describe where something happens requires three "coordinates." Suppose that I want to give directions to an invisible balloon floating in an empty room. I could say, for example, "Starting in a certain corner, go five feet along a certain wall (one dimension), four feet directly out from the wall (second dimension), and three feet up from the floor (third dimension)." Every possibility exists in three dimensions and has a time.

If the wave function represents possibilities associated with two different particles, then that wave function exists in six dimensions, three for each particle. If the wave function represents the possibilities associated with twelve particles, then that wave function exists in thirty-six dimensions!*

This is impossible to visualize since our experience is limited to three dimensions. Nontheless, this is the mathematics of the situation.

The point to think about is that when we make a measurement in a quantum mechanical experiment—when the observed system interacts with the observing system—we reduce a multidimensional reality to a three-dimensional reality compatible with our experience.

If we calculate a wave function for possible photon detection at four different points, that wave function is a mathematical reality in which four different happenings exist simultaneously in twelve dimensions. In principle, we can calculate a wave function representing an infinite number of events happening at the same time in an infinite number of dimensions. No matter how complex the wave function,

---

*The state of a system containing n particles is represented at each time by a wave function in a $3n$ dimensional space. If we make an observation on each of the n particles the wave function is reduced to a special form—to a product of n wave functions each of which is in a three-dimensional space. Thus the number of dimensions in the wave function is determined by the number of particles in the system.

however, as soon as we make a measurement, we reduce it to a form compatible with three-dimensional reality, which is the only form of experiential reality, instant by instant, normally available to us.

Now we come to the question, "When, exactly, does the wave function collapse?" When do all of the possibilities that are developing for the observed system, except one, vanish?

Up to now, we have said that the collapse occurs when somebody looks at the observed system. This is only one point of view. Another opinion (any discussion about this question is opinion) is that the wave function collapses when *I* look at the observed system. Still another opinion is that the wave function collapses when any measurement is made, even by an instrument. According to this view, it is not important whether we are there to see it or not.

Suppose for the moment that there are no human experimenters involved in our experiment. It is entirely automatic. A light source emits a photon. The wave function of the photon contains the possibility that the photon will pass through slit one and detector one will fire, and also the possibility that the photon will pass through slit two and detector two will fire.

Now suppose that detector two registers a photon.

According to classical physics, the light source emitted a real particle, a photon, and it traveled from the light source to the slit where detector two recorded it. Although we did not know its location while it was in transit, we could have determined it, if we had known how.

According to quantum mechanics, this is not so. No real particle called a photon traveled between the light source and the screen. There was no photon until one actualized at slit two. Until then, there was only a wave function. In other words, until then, all that existed were tendencies for a photon to actualize either at slit one or at slit two.

From the classical point of view, a real photon travels between the light source and the screen. The odds are 50–50 that it will go to slit one and 50–50 that it will go to slit two. From the point of view of quantum mechanics, there is no photon until a detector fires. There is only a developing potentiality in which a photon goes to slit one *and* to slit two. This is Heisenberg's "strange kind of physical reality just in the middle between possibility and reality."[5]

It is difficult to make this sound less vague. The translation from mathematics to English entails a loss of precision but that is not the problem. We can experience a more clearly defined picture of this phenomenon by learning enough mathematics to follow the development of the Schrödinger wave equation. Unfortunately, clarifying the picture only helps to boggle the mind.

The real problem is that we are used to looking at the world simply. We are accustomed to believing that something is there or it is not there. Whether we look at it or not, it is either there or it is not there. Our experience tells us that the physical world is solid, real, and independent of us. Quantum mechanics says, simply, that this is not so.

Suppose that a technician, not knowing that our experiment is automatic, enters the room to see which detector has recorded a photon. When he looks at the observing system (the detectors), there are two things that he can see. The first possibility is that detector one has recorded the photon, and the second possibility is that detector two has recorded the photon. The wave function of the observing system (which now is the technician), therefore, has two humps in it, one for each possibility.

Until the technician looks at the detectors, quantum mechanically speaking, both situations in some way exist. As soon as he sees that detector two has fired, however, the possibility that detector one has fired vanishes. That part of the wave function of the measuring system collapses, and the reality of the technician is that detector two has recorded a photon. In other words, the observing system of the experiment, the detectors, has become the observed system in relation to the technician.

Now suppose that the supervising physicist enters the room to check on the technician. He wants to see what the technician has learned about the detectors. In this regard, there are two possibilities.

One is that the technician has seen that detector one has recorded a photon, and the other is that the technician has seen that detector two has recorded a photon, and so on.*

The division of the wave function into two humps, each one representing a possibility, has progressed from photon to detectors to technician to supervisor. This proliferation of possibilities is the type of development governed by the Schrödinger wave equation.

Without perception, the universe continues, via the Schrödinger equation, to generate an endless profusion of possibilities. The effect of perception, however, is immediate and dramatic. All of the wave function representing the observed system collapses, except one part, which actualizes into reality. No one knows what causes a particular possibility to actualize and the rest to vanish. The only law governing this phenomenon is statistical. In other words, it is up to chance.

The division into two parts of the wave function of the photon, detectors, technician, supervisor, etc., is known as the "Problem of Measurement" (or, sometimes, "The Theory of Measurement").† If there were twenty-five possibilities in the wave function of the photon, the wave function of the measuring system, technician, and supervisor similarly would have twenty-five separate humps, until a perception is made and the wave function collapses. From photon to detectors to technician to supervisor we could continue until we include the entire universe. *Who is looking at the universe?* Put another way, *How is the universe being actualized?*

The answer comes full circle. *We* are actualizing the universe. Since we are part of the universe, that makes the universe (and us) self-actualizing.

---

*To see the conciseness of mathematical expression, consider that the entire process described in the Theory of Measurement, from photon (system, S) to detectors (measuring device, M) to technician (observer, O) can be represented mathematically by one "sentence":

$$(\Psi_S^1 + \Psi_S^2) \otimes \Psi_M \otimes \Psi_O \rightarrow \Sigma(\overline{\Psi}_S^1 \otimes \overline{\Psi}_M^1 \otimes \overline{\Psi}_O^1) + \Sigma(\overline{\Psi}_S^2 \otimes \overline{\Psi}_M^2 \otimes \overline{\Psi}_O^2)$$

†The Theory of Measurement presented here is essentially from John von Neumann's 1932 discussion.

This line of thought is similar to some aspects of Buddhist psychology. In addition, it could become one of many important contributions of physics to future models of consciousness.

The Copenhagen Interpretation of Quantum Mechanics says that it is unnecessary to "peer behind the scenes to see what is really happening as long as quantum mechanics works (correlates experience correctly) in all possible experimental situations. It is not necessary to know how light can manifest itself both as particles and waves. It is enough to know that it does and to be able to use this phenomenon to predict probabilities. In other words, the wave and particle characteristics of light are unified by quantum mechanics, but at a price. There is no description of reality.

All attempts to describe "reality" are relegated to the realm of metaphysical speculation.* However, this does not mean that physicists do not speculate. Many do, in particular Henry Stapp, and their reasoning goes like this.

The fundamental theoretical quantity in quantum mechanics is the wave function. The wave function is a dynamic (it changes as time progresses) description of possible occurrences. But what does the wave function describe, really? According to western thought, the world has only two essential aspects, one of which is matter-like and the other of which is idea-like.

The matter-like aspect is associated with the external world, most of which is conceived to be made of inanimate stuff that is hard and unresponsive, like rocks, pavement, metal, etc. The idea-like aspect is our subjective experience. Reconciling these two has been a central theme of religion through history. The philosophies which champion these aspects are Materialism (the world is matter-like, regardless of our impressions) and Idealism (reality is idea-like, regardless of

---

*The wave function is the physicist's description of reality. At issue is the interpretation of the wave function and whether it is the best possible description (or simply the only one that fits the language used by physicists).

appearances). The question is, which one of these aspects does the wave function represent?

The answer, according to the orthodox view of quantum mechanics elucidated by Stapp, is that the wave function represents something that partakes of *both* idea-like and matter-like characteristics.*

For example, when the observed system as represented by the wave function propagates in isolation between the region of preparation and the region of measurement, it develops according to a strictly deterministic law (the Schrödinger wave equation). Temporal development in accordance with a causal law is a matter-like characteristic. Therefore, whatever the wave function represents, that something has a matter-like aspect.

However, when the observed system as represented by the wave function interacts with the observing system (when we make a measurement), it abruptly leaps to a new state. These "Quantum Leap" type transitions are idea-like characteristics. Ideas (like our knowledge about something) can and do change discontinuously. Therefore, whatever the wave function represents, that something also has an idea-like aspect.

The wave function, strictly speaking, represents an observed system in a quantum mechanical experiment. In more general terms, it describes physical reality at the most fundamental level (the sub-

---

*The wave function, since it is a tool for our understanding of nature, is something in our thoughts. It represents certain *specifications* of certain physical systems. Specifications are objective in the sense that scientists and technicians can agree on them. However, specifications do not exist apart from thought. Also, any given physical system satisfies many sets of specifications, and many physical systems can satisfy one set of specifications. All of these characteristics are idea-like and, to that extent, that which is represented by the wave function is idea-like, even though it is objective.

However, these specifications are transcribed into wave functions that develop according to a determined law (the Schrödinger wave equation). This is a matter-like aspect. The thing that develops describes only probabilities. Probabilities can be thought to describe either things that exist apart from thought, or things that exist only within thought. Thus that which the wave function represents has both idea-like and matter-like characteristics.

atomic) that physicists have been able to probe. In fact, according to quantum mechanics, the wave function is a *complete* description of physical reality at that level. Most physicists believe that a description of the substructure underlying experience more complete than the wave function is not possible.

"Wait a minute!" says Jim de Wit (where did he come from?). "The description contained in the wave function consists of coordinates (three, six, nine, etc.) and a time (page 84). How can that be a complete description of reality? Imagine how I felt when my girlfriend ran off to Mexico with a gypsy. Where does *that* show up in a wave function?"

It doesn't. The "complete description" that quantum theory claims the wave function to be is a description of *physical* reality (as in *physics*). No matter what we are feeling, or thinking about, or looking at, the wave function describes as completely as possible where and when we are doing it.

Since the wave function is thought to be a complete description of physical reality and since that which the wave function describes is idea-like as well as matter-like, then physical reality must be both idea-like and matter-like. In other words, the world cannot be as it appears. Incredible as it sounds, this is the conclusion of the orthodox view of quantum mechanics. The physical world *appears* to be completely substantive (made of "stuff"). Nonetheless, if it has an idea-like aspect, the physical world is not substantive in the usual sense of the word (one hundred percent matter, zero percent idea). According to Stapp:

> If the attitude of quantum mechanics is correct, in the strong sense that a description of the substructure underlying experience more complete than the one it provides is not possible, then there is no substantive physical world, in the usual sense of this term. The conclusion here is not the weak conclusion that there *may* not be a substantive physical world but rather that there definitely is not a substantive physical world.[6]

This does not mean that the world is completely idea-like. The Copenhagen Interpretation of Quantum Mechanics does not go so far as to say what reality is "really like behind the scenes," but it does say that it is not like it appears. It says that what we perceive to be physical reality is actually our cognitive construction of it. This cognitive construction may appear to be substantive, but the Copenhagen Interpretation of Quantum Mechanics leads directly to the conclusion that the physical world itself is not.

This claim at first appears so preposterous and remote from experience that our inclination is to discard it as the foolish product of cloistered intellectuals. However, there are several good reasons why we should not be so hasty. The first reason is that quantum mechanics is a logically consistent system. It is self-consistent and it also is consistent with all known experiments.

Second, the experimental evidence itself is incompatible with our ordinary ideas about reality.

Third, physicists are not the only people who view the world this way. They are only the newest members of a sizable group; most Hindus and Buddhists also hold similar views.

Therefore, it is evident that even physicists who disclaim metaphysics have difficulty avoiding it. Now we come to those physicists who have jumped feet first into describing "reality."

So far our discussions have been based on the Copenhagen Interpretation of Quantum Mechanics. The unavoidable flaw in this interpretation is the Problem of Measurement. Some type of detection by an observing system is required to collapse the wave function of the observed system into a physical reality, otherwise the "observed system" does not physically exist except as an endlessly proliferating number of possibilities generated in accordance with the Schrödinger wave equation.

The theory proposed by Hugh Everett, John Wheeler, and Neill Graham solves this problem in the simplest way possible.[7] It claims that the wave function is a real thing, all of the possibilities that it represents are real, *and they all happen*. The orthodox interpretation of quantum mechanics is that only one of the possibilities contained

in the wave function of an observed system actualizes, and the rest
vanish. The Everett-Wheeler-Graham theory says that they *all* actualize, but in different worlds that coexist with ours!

Let's go back to the double-slit experiment again. A light source
emits a photon. The photon can pass through slit one or through slit
two. A detector is placed at slit one and at slit two. Now we add a new
experimental procedure. If the photon goes through slit one, I run
upstairs. If the photon goes through slit two, I run downstairs. Therefore, one possible occurrence is that the photon goes through slit one,
detector one fires, and I run up the stairs. The second possible occurrence is that the photon goes through slit two, detector two fires, and
I run down the stairs.

According to the Copenhagen Interpretation, these two possibilities are mutually exclusive because it is not possible for me to run
upstairs and to run downstairs at the same time.

According to the Everett-Wheeler-Graham theory, at the
moment the wave function "collapses," the universe splits into two
worlds. In one of them I run up the stairs and in the other I run down
the stairs. There are two distinct editions of me. Each one of them is
doing something different, and each one of them is unaware of the
other. Nor will their (our) paths ever cross since the two worlds into
which the original one split are forever separate branches of reality.

In other words, according to the Copenhagen Interpretation of
Quantum Mechanics, the development of the Schrödinger wave equation generates an endlessly proliferating number of possibilities.
According to the Everett-Wheeler-Graham theory, the development
of the Schrödinger wave equation generates an endlessly proliferating
number of *different branches of reality!* This theory is called, appropriately, the Many Worlds Interpretation of Quantum Mechanics.

The theoretical advantage of the Many Worlds Interpretation is
that it does not require an "external observer" to "collapse" one of
the possibilities contained in a wave function into physical reality.
According to the Many Worlds theory, wave functions do not collapse, they just keep splitting as they develop according to the Schrödinger wave equation. When a consciousness happens to be present at

such a split, it splits also, one part of it associating with one branch of reality and the other part(s) of it associating with the other branch(es) of reality. However, each branch of reality is experientially inaccessable to the other(s), and a consciousness in any one branch will consider that branch to be the entirety of reality. Therefore, the role of consciousness, which was central to the Copenhagen Interpretation (if consciousness is associated with an act of measurement), is incidental to the Many Worlds theory.

However, the Many Worlds description of the structure of the relationship between the various branches of physical reality sounds like a quantitative version of a mystical vision of unity. Every state of a subsystem of a composite system is uniquely correlated to the states of the remaining subsystems which constitute the whole of which it is a part. (A "composite system," in this case, means a combination of both the observed system and the observing system. In other words, every state of the observed system is correlated to a particular state of the observing system).

Said another way, the Many Worlds theory defines any particular branch of reality which might "actualize" to us as a result of an interaction of an observed system and an observing system as merely one way of decomposing the wave function which represents them both. According to this theory, all of the other states which "could have" resulted from the same interaction *did happen,* but in other branches of reality. Each of these branches of reality are *real,* and, together, they constitute all the different ways in which we can decompose the universal wave function.

In this way, the Problem of Measurement is no longer a problem. The problem of measurement, ultimately, was, "Who is looking at the universe?" The Many Worlds theory says that it is not necessary to collapse a wave function to actualize the universe. All of the mutually exclusive possibilities contained within the wave function of an observed system that (according to the Copenhagen Interpretation) do not actualize when the wave function "collapses" actually *do* actualize, but not in this branch of the universe. In our experiment, for example, one of the possibilities contained in the wave function actualizes in this branch of the universe (I run up the stairs). The other possibility contained in the wave function (I run down the stairs) also actualizes, but in a different branch of reality. In this branch of reality I run up the stairs. In another branch of reality I run down the stairs. Neither "I" knows the other. Both "I"s believe that their branch of the universe is the entirety of reality.

The Many Worlds theory says that there is one universe and that its wave function represents all of the ways that it can be decomposed into different possible realities. We are all together here in a big box and it is not necessary to look at the box from the outside to actualize it.

In this regard, the Many Worlds theory is especially interesting because Einstein's general theory of relativity shows that our universe might *be* something like a large closed box and, if this is so, it is never possible to get "outside" of it.*

"Schrödinger's Cat" sums up the differences between classical physics, the Copenhagen Interpretation of Quantum Mechanics, and the Many Worlds Interpretation of Quantum Mechanics. "Schrödinger's Cat" is a dilemma posed long ago by the famous discoverer of the Schrödinger wave equation:

A cat is placed inside a box. Inside the box is a device which can release a gas, instantly killing the cat. A random event (the radioactive decay of an atom) determines whether the gas is released or not. There is no way of knowing, outside of looking into the box, what happens inside it. The box is sealed and the experiment is activated. A moment later, the gas either has been released or has not been released. The question is, without looking, what has happened inside the box. (This is reminiscent of Einstein's unopenable watch.)

According to classical physics, the cat is either dead or it is not

---

*"How is one to apply the conventional formulation of quantum mechanics to the space-time geometry itself? The issue becomes especially acute in the case of a closed universe. There is no place to stand outside the system to observe it."—Hugh Everett III (*Reviews of Modern Physics*, 29, 3, 1957, 455).

dead. All that we have to do is to open the box and see which is the case. According to quantum mechanics, the situation is not so simple.

The Copenhagen Interpretation of Quantum Mechanics says that the cat is in a kind of limbo represented by a wave function which contains the possibility that the cat is dead and also the possibility that the cat is alive.* When we look in the box, and not before, one of these possibilities actualizes and the other vanishes. This is known as the collapse of the wave function because the hump in the wave function representing the possibility that did not occur, collapses. It is necessary to look into the box before either possibility can occur. Until then, there is only a wave function.

Of course, this does not make sense. Experience tells us that a cat is what we put into the box and a cat is still what is inside the box, not a wave function. The only question is whether the cat is a live cat or a dead cat. But a cat *is there* whether we look at it or not. If we take a vacation before we look inside the box, it makes no difference as far as the cat is concerned. Its fate was decided at the beginning of the experiment.

This commonsense view is also the view of classical physics. According to classical physics, we get to know something by observing it. According to quantum mechanics, it *isn't there* until we do observe it! Therefore, the fate of the cat is not determined until we look inside the box.

The Many Worlds Interpretation of Quantum Mechanics and the Copenhagen Interpretation of Quantum Mechanics agree that the fate of the cat is not determined for us until we look inside the box. What happens after we look inside the box, however, depends upon which interpretation we choose to follow. According to the Copenhagen Interpretation, at the instant that we look inside the box, one of the possibilities contained in the wave function representing the cat

---

*In practice, it is not clear that a macroscopic object such as a cat actually can be represented by a wave function due to the dominating influence of thermodynamically irreversible processes. Even so, Schrödinger's cat long has illustrated to physics students the psychedelic aspects of quantum mechanics.

actualizes and the other possibility vanishes. The cat is either dead or alive.

According to the Many Worlds Interpretation, at the instant that the atom decays (or doesn't decay, depending upon which branch of reality we are talking about), the world splits into two branches, each with a different edition of the cat. The wave function representing the cat does not collapse. The cat is both dead *and* alive. At the instant that we look into the box, our wave function also splits into two branches, one associated with the branch of reality in which the cat is dead and one associated with the branch of reality in which the cat is alive. Neither consciousness is aware of the other.

In short, classical physics says that there is one world, it is as it appears, and this is it. Quantum physics allows us to entertain the possibility that this is not so. The Copenhagen Interpretation of Quantum Mechanics eschews a description of what the world is "really like," but concludes that whatever it is like, it is not substantive in the usual sense. The Many Worlds Interpretation of Quantum Mechanics says that different editions of us live in many worlds simultaneously, an uncountable number of them, and all of them are real. There are even more interpretations of quantum mechanics, but all of them are weird in some way.

Quantum physics is stranger than science fiction.

Quantum mechanics is a theory and a procedure dealing with subatomic phenomena. Subatomic phenomena, in general, are inaccessible to all but those with access to elaborate (and expensive) facilities. Even at the most expensive and elaborate facilities, however, we can see only the effects of subatomic phenomena. The subatomic realm is beyond the limits of sensory perception.* It is also beyond the limits of rational understanding. Of course, we have rational theories about it, but "rational" has been stretched to include what formerly was nonsense, or, at best, paradox.

*The dark-adapted eye can detect single photons. All of the other subatomic particles must be detected indirectly.

The world that we live in, the world of freeways, bathtubs, and other people, seems as remote as it can be from wave functions and interference. In short, the metaphysics of quantum mechanics is based upon an unsubstantiated leap from the microscopic to the macroscopic. Can we apply these implications of subatomic research to the world at large?

No, not if we have to provide a mathematical proof in each instance. But what is a proof? A proof only proves that we are playing by the rules. (We make the rules, anyway.) The rules, in this case, are that what we propose about the nature of physical reality (1) be logically consistent, and (2) that it correspond to experience. There is nothing in the rules that says that what we propose has to be anything like "reality." Physics is a self-consistent explanation of experience. It is in order to satisfy the self-consistency requirement of physics that proofs become important.

The New Testament presents a different point of view. Christ, following His resurrection, proved to Thomas (who became the proverbial "Doubting Thomas") that He really was He, risen from the dead, by showing Thomas His wounds. At the same time, however, Christ bestowed His special favor on those who believed Him *without proof*.

Acceptance without proof is the fundamental characteristic of western religion. Rejection without proof is the fundamental characteristic of western science. In other words, religion has become a matter of the heart and science has become a matter of the mind. This regrettable state of affairs does not reflect the fact that, physiologically, one cannot exist without the other. Everybody needs both. Mind and heart are only different aspects of *us*.

Who, then, is right? Should disciples believe without proof? Should scientists insist on it? Is the world without substance? Is it real, but divided and dividing into countless branches?

The Wu Li Masters know that "science" and "religion" are only dances, and that those who follow them are dancers. The dancers may claim to follow "truth" or claim to seek "reality," but the Wu Li Masters know better. They know that the true love of all dancers is dancing.

Part One

MY WAY

1

✺

The Role of "I"

In the days before Copernicus discovered that the earth revolves around the sun, the common belief was that the sun, along with the rest of the universe, revolved around the earth. The earth was the fixed center of everything. At a still earlier time in India, this geocentric position was given to people. That is, each person, psychologically speaking, was recognized as being the center of the universe. Although this sounds like an egotistical point of view, it was not since *every* person was recognized as a divine manifestation.

A beautiful Hindu painting shows Lord Krishna dancing in the moonlight on the bank of the Yamuna. He moves in the center of a circle of fair Vraja women. They are all in love with Krishna and they are dancing with him. Krishna is dancing with all of the souls of the world—man is dancing with himself. To dance with god, the creator of all things, is to dance with ourselves. This is a recurrent theme of eastern literature.

This is also the direction toward which the new physics, quantum mechanics and relativity, seems to point. From the revolutionary concepts of relativity and the logic-defying paradoxes of quantum mechanics an ancient paradigm is emerging. In vague form, we begin to glimpse a conceptual framework in which each of us shares a paternity in the creation of physical reality. Our old self-image as impotent bystander, one who sees but does not affect, is dissolving.

We are watching perhaps the most engaging act in our history. Amid the powerful purr of particle accelerators, the click of computer printouts, and dancing instrument gauges, the "old science" that has given us so much, including our sense of helplessness before the faceless forces of bigness, is undermining its own foundations.

With the awesome authority that we have given it, science is telling us that our faith has been misplaced. It appears that we have attempted the impossible, to disown our part in the universe. We have tried to do this by relinquishing our authority to the Scientists. To the Scientists we gave the responsibility of probing the mysteries of creation, change, and death. To us we gave the everyday routine of mindless living.

The Scientists readily assumed their task. We readily assumed ours, which was to play a role of impotence before the ever-increasing complexity of "modern science" and the ever-spreading specialization of modern technology.

Now, after three centuries, the Scientists have returned with their discoveries. They are as perplexed as we are (those of them who have given thought to what is happening).

"We are not sure," they tell us, "but we have accumulated evidence which indicates that the key to understanding the universe is *you*."

This is not only different from the way that we have looked at the world for three hundred years, it is *opposite*. The distinction between the "in here" and the "out there" upon which science was founded, is becoming blurred. This is a puzzling state of affairs. Scientists, using the "in here—out there" distinction, have discovered that the "in here—out there" distinction may not exist! What is "out there" apparently depends, in a rigorous mathematical sense as well as a philosophical one, upon what we decide "in here."

The new physics tells us that an observer cannot observe without altering what he sees. Observer and observed are interrelated in a real and fundamental sense. The exact nature of this interrelation is not clear, but there is a growing body of evidence that the distinction between the "in here" and the "out there" is illusion.

The conceptual framework of quantum mechanics, supported by massive volumes of experimental data, forces contemporary physicists to express themselves in a manner that sounds, even to the uninitiated, like the language of mystics.

Access to the physical world is through experience. The common denominator of all experiences is the "I" that does the experiencing. In short, what we experience is not external reality, but our *interaction* with it. This is a fundamental assumption of "complementarity."

Complementarity is the concept developed by Niels Bohr to explain the wave-particle duality of light. No one has thought of a better one yet. Wave-like characteristics and particle-like characteristics, the theory goes, are mutually exclusive, or complementary aspects of light. Although one of them always excludes the other, *both* of them are necessary to understand light. One of them always excludes the other because light, or anything else, cannot be both wave-like and particle-like at the same time.*

How can mutually exclusive wave-like and particle-like behaviors both be properties of one and the same light? They are not properties of light. They are properties of our *interaction* with light. Depending upon our choice of experiment, we can cause light to manifest either particle-like properties or wave-like properties. If we choose to demonstrate the wave-like characteristics of light, we can perform the double-slit experiment which produces interference. If we choose to demonstrate the particle-like characteristics of light, we can perform an experiment which illustrates the photoelectric effect. We can cause light to manifest both wave-like properties and particle-like properties by performing Arthur Compton's famous experiment.

In 1923, Compton played the world's first game of billiards with

---

*Individual events are always particle-like; wave behavior is detected as a statistical pattern, i.e., interference. However, in the words of Paul Dirac (another founder of quantum mechanics) even a single subatomic particle "interferes with itself." How a single subatomic particle, like an electron, for example, can "interfere with itself" is the basic quantum paradox.

subatomic particles, and, in the process, confirmed Einstein's seventeen-year-old photon theory of light. His experiment was not conceptually difficult. He simply fired x-rays, which everybody knows are waves, at electrons. To the surprise of most people, the x-rays bounced off the electrons as if they (the x-rays) were particles! For example, the x-rays which struck the electrons glancing blows were deflected only slightly from their paths. They did not lose much energy in the collision. However, those x-rays which collided more nearly head-on with electrons were deflected sharply. These x-rays lost a considerable amount of their kinetic energy (the energy of motion) in the collision.

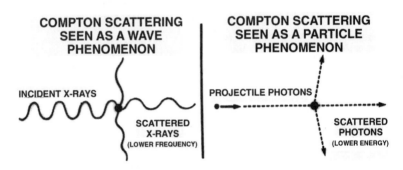

COMPTON SCATTERING SEEN AS A WAVE PHENOMENON

INCIDENT X-RAYS

SCATTERED X-RAYS (LOWER FREQUENCY)

COMPTON SCATTERING SEEN AS A PARTICLE PHENOMENON

PROJECTILE PHOTONS

SCATTERED PHOTONS (LOWER ENERGY)

Compton could tell just how much energy the deflected x-rays lost by measuring their frequencies before and after the collision. The frequencies of those x-rays involved in near head-on collisions were noticeably lower after the collision than before it. This meant that they had less energy after the collision than they had before the collision. Compton's x-rays were impacting with electrons exactly the way that billiard balls impact with other billiard balls.

Compton's discovery was intimately related to quantum theory. Compton could not have revealed the particle-like behavior of x-rays if Planck had not discovered his fundamental rule that higher frequency means higher energy. This rule permitted Compton to prove that the x-rays in his experiment lost energy in a particle-like collision (because their frequencies were lower after the collision than before the collision).

The conceptual paradox in Compton's experiment shows how

deeply the wave-particle duality is embedded in quantum mechanics. Compton proved that electromagnetic radiations, like x-rays, have particle-like characteristics by measuring their frequencies! Of course, "particles" don't have frequencies. Only waves have frequencies. The phenomenon which Compton discovered is called Compton scattering, in honor of what happens to the x-rays.

In short, we can demonstrate that light is particle-like with the photoelectric effect, that it is wave-like with the double-slit experiment, and that it is both particle-like and wave-like with Compton scattering. Both of these complementary aspects of light (wave and particle) are necessary to understand the nature of light. It is meaningless to ask which one of them, alone, is the way light really is. Light behaves like waves or like particles depending upon which experiment we perform.

The "we" that does the experimenting is the common link that connects light as particles and light as waves. The wave-like behavior that we observe in the double-slit experiment is not a property of light, it is a property of our interaction with light. Similarly, the particle-like characteristics that we observe in the photoelectric effect are not a property of light. They, too, are a property of our interaction with light. Wave-like behavior and particle-like behavior are properties of *interactions.*

Since particle-like behavior and wave-like behavior are the only properties that we ascribe to light, and since these properties now are recognized to belong (if complementarity is correct) not to light itself, but to our interaction with light, then it appears that light has no properties independent of us! To say that something has no properties is the same as saying that it does not exist. The next step is this logic is inescapable. Without us, light does not exist.

Transferring the properties that we usually ascribe to light to our interaction with light deprives light of an independent existence. Without us, or by implication, anything else to interact with, light does not exist. This remarkable conclusion is only half the story. The other half is that, in a similar manner, without light, or, by implication, anything else to interact with, *we do not exist!* As Bohr himself put it:

> . . . an independent reality in the ordinary physical sense can be ascribed neither to the phenomena nor to the agencies of observation.[1]

By "agencies of observation," he may have been referring to instruments, not people, but philosophically, complementarity leads to the conclusion that the world consists not of things, but of interactions. Properties belong to interactions, not to independently existing things, like "light." This is the way that Bohr solved the wave-particle duality of light. The philosophical implications of complementarity became even more pronounced with the discovery that the wave-particle duality is a characteristic of *everything*.

When we left off telling the story of quantum mechanics, the tale had progressed as follows: In 1900, Max Planck, studying black-body radiation, discovered that energy is absorbed and emitted in chunks, which he called quanta. Until that time, radiated energy, like light, was thought to be wave-like. This was because Thomas Young, in 1803, showed that light produces interference (the double-slit experiment), and only waves can do that.

Einstein, stimulated by Planck's discovery of quanta, used the photoelectric effect to illustrate his theory that not only are the processes of energy absorption and emission quantized, but that *energy itself* comes in packages of certain sizes. Thus physicists were confronted with two sets of experiments (repeatable experiences) each of which seemed to disprove the other. This is the famous wave-particle duality which is fundamental to quantum mechanics.

While physicists were trying to explain how waves can be particles, a young French prince, Louis de Broglie, dropped a bomb which demolished what was left of the classical view. Not only are waves particles, he proposed, but particles are also waves!

De Broglie's idea (which was contained in his doctoral thesis) was that matter has waves which "correspond" to it. The idea was more than philosophical speculation. It was also mathematical speculation. Using the simple equations of Planck and Einstein, de Broglie formulated a simple equation of his own.* It determines the wavelength of the "matter waves" that "correspond" to matter. It says simply that the greater the momentum of a particle, the shorter is the length of its associated wave.

This explains why matter waves are not evident in the macroscopic world. De Broglie's equation tells us that the matter waves corresponding to even the smallest object that we can see are so incredibly small compared to the size of the object that their effect is negligible. However, when we get down to something as small as a subatomic particle, like an electron, the size of the electron itself is smaller than the length of its associated wave!

Under these circumstances, the wave-like behavior of matter should be clearly evident, and matter should behave differently than "matter" as we are used to thinking of it. This is exactly what happens.

Only two years after de Broglie presented this hypothesis, an experimenter named Clinton Davisson, working with his assistant, Lester Germer, at the Bell Telephone Laboratories, verified it experimentally. Both Davisson and de Broglie got Nobel Prizes, and physicists were left to explain not only how waves can be particles, but also how particles can be waves.

The famous Davisson-Germer experiment, which was done by accident, showed electrons reflecting off a crystal surface in a manner that could be explained only if the electrons were waves. But, of course, electrons are particles.

Today, electron diffraction, an apparent contradiction in terms, is a common phenomenon. When a beam of electrons is sent through tiny openings, like the spaces between the atoms in a metal foil, which are as small or smaller than the wavelengths of the electrons (isn't this ridiculous—"particles" *don't have* wavelengths!), the beam diffracts

*Planck's equation: $E = h\nu$. Einstein's equation: $E = mc^2$. De Broglie's equation: $\lambda = h/mv$.

exactly the way a beam of light diffracts. Although, classically speaking, it can't happen, here is a picture of it.

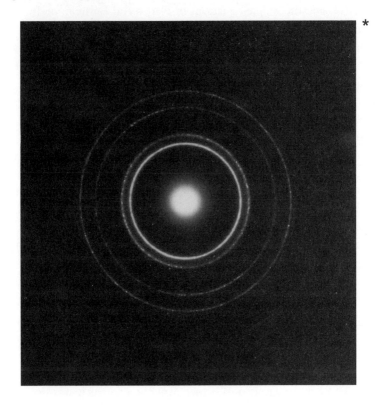

*

It was disconcerting enough when light, which is made of waves, began to behave like particles, but when electrons, which *are* particles, began to behave like waves, the plot became unbearably thick.

The unfolding of quantum mechanics was (and still is) a drama of high suspense. Werner Heisenberg wrote:

*As you hold this photograph in front of you, the beam of electrons (the "transmitted beam") is coming directly toward you out of the large white spot in the center. Also located in the white spot is the diffracting material (in this case, the electron beam is being diffracted by small grains of gold, i.e., the beam is being directed through a thin polycrystalline gold foil). The rings on the photo-

> I remember discussions with Bohr [in 1927] which went through many hours till very late at night and ended almost in despair; and when at the end of the discussion I went alone for a walk in the neighboring park I repeated to myself again and again the question: Can nature possibly be as absurd as it seemed to us in these atomic experiments.[2]

Subsequent experiments were to reveal that not only subatomic particles, but atoms and molecules as well have associated matter waves. The title of Donald Hughes's pioneer book, *Neutron Optics,* provides eloquent testimony of the merger between waves and particles to which Prince de Broglie's doctoral thesis gave birth. Theoretically, in fact, *everything* has a wavelength—baseballs, automobiles, and even people—although their wavelengths are so small that they are not noticeable.

De Broglie himself was not very helpful in explaining his theory. It predicted what the Davisson-Germer experiment proved: that matter, like electrons, has a wave-like aspect. His equation even foretold the wavelength of these waves. Nonetheless, no one knew what these waves actually were (no one does yet). De Broglie called them waves which "correspond" to matter, but he did not explain what "correspond" meant.

Is it possible for a physicist to predict something, calculate equations which describe it, and still not know what he is talking about?

Yes. As Bertrand Russell put it:

> Mathematics may be defined as the subject in which we never know what we are talking about, nor whether what we are saying is true.[3]

This is why the physicists at Copenhagen decided to accept quantum mechanics as a complete theory even though it gives no explana-

---

graph mark the places where the diffracted electron beams struck the film which was placed on the opposite side of the gold foil from the electron source. The white spot in the center of the photograph was caused by undiffracted electrons in the transmitted beam passing through the gold foil and striking the film directly.

tion of what the world is "really like," and even though it predicts probabilities and not actual events. They accepted quantum
mechanics as a complete theory because quantum mechanics correctly
correlates experience. Quantum mechanics, and, according to the
pragmatists, all science, is the study of correlations between experiences. De Broglie's equation correctly correlates experiences.

De Broglie merged the wave-particle paradox which came to
light (hissss) through the genius of Thomas Young (double-slit experiment) and Albert Einstein (photon theory). In other words, he connected the two most revolutionary phenomena of physics, the
quantum nature of energy and the wave-particle duality.

De Broglie presented his matter-wave theory in 1924. During
the next three years quantum mechanics crystallized into what it
essentially is today. The world of Newtonian physics, simple mental
pictures, and common sense disappeared. A new physics took form
with an originality and force that left the mind reeling.

After de Broglie's matter waves came the Schrödinger wave
equation.

De Broglie's matter waves seemed to Erwin Schrödinger, the Vien-
nese physicist, a much more natural way of looking at atomic phenomena than Bohr's planetary model of the atom. Bohr's model of hard,
spherical electrons revolving around a nucleus at specific levels and
emitting photons by jumping from one level to another explained the
color spectrum of simple atoms, but it said nothing about why each
shell contains only a certain number of electrons, no more and no less.
It also did not explain *how* the electrons do their jumping (for example, what is happening to them between shells).*

Stimulated by de Broglie's discovery, Schrödinger hypothesized
that electrons are not spherical objects, but *patterns of standing waves.*

Standing waves are familiar phenomena to anyone who has

---

*Accurately speaking, Schrödinger's theory does not explain The Jump,
either. In fact, Schrödinger did not like the idea of a "jump."

played with a clothesline. Suppose that we tie one end of a rope to a pole, and then pull it tight. On this rope there are no waves at all, either standing or traveling. Now suppose that we flick our wrist sharply downward and then upward. A hump appears in the rope and travels down the rope to the pole where it turns upside down and returns to our hand. This traveling hump (figure A) is a traveling wave. By sending a series of humps down the rope, we can set up the patterns of standing waves shown below, and more that are not shown.

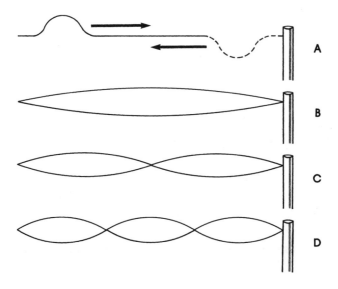

The simplest of these is the pattern shown in figure B. This pattern is formed by the superposition of two traveling waves, a direct one and a reflected one traveling in the opposite direction. It is the pattern, not the rope, which does not move. The widest point in the standing wave remains "stationary," and so do the points at the ends of the standing wave. These points are called nodes. There are two of them in the simplest standing pattern, one at our hand and one at the pole where the rope is attached. These stationary patterns, superpositions of traveling waves, are called standing waves.

No matter how long or short our rope is, there can be only a whole number of standing waves on it. That is, it can have a pattern of one standing wave, or a pattern of two standing waves, or a pattern of three, four, five, and so on, standing waves but it can never have a pattern of one and one half standing waves, or a pattern of two and one fourth standing waves. The standing waves must divide the rope evenly into whole sections. Another way to say this is that we can increase or decrease the number of standing waves on a rope only by a whole number of them. This means that the only way that the number of standing waves on a rope can increase or decrease is *discontinuously* (that word, again!).

Furthermore, standing waves on a rope cannot be just any size. They always will be restricted to those lengths which divide the rope evenly. The actual size of the waves depends upon how long the rope is, but no matter what length the rope, there will be only certain lengths which divide it evenly.

All of this was old stuff in 1925. Plucking a guitar string establishes patterns of standing waves on it. Blowing air into an organ pipe creates standing wave patterns in it. What was new was Schrödinger's realization that *standing waves are "quantized" the same way that atomic phenomena are!* In fact, Schrödinger proposed that electrons *are* standing waves.

In retrospect, this is not as fantastic as it first sounds. At the time, however, it was a stroke of genius. Picture an electron in orbit around a nucleus. Each time the electron completes a journey around the nucleus, it travels a certain distance. That distance is a certain length, like our rope was a certain length. Similarly, only a whole number of standing waves, never a fraction of one, can form in this length. (Length of what is an unanswered question.)

Schrödinger proposed that each of these standing waves is an electron! In other words, he proposed that electrons are the segments of vibrations bounded by the nodes. A drawing of this is on the next page.

So far, we have talked about standing waves on a line, like a clothesline or a guitar string, but standing waves also occur in other mediums, like water. Suppose that we throw a rock into a round pool. Waves radiate from its point of entry. These waves are reflected, sometimes more than once, off different sides of the pool. When the reflected traveling waves interfere with each other they create a complex pattern of standing waves which is our old friend, interference.

Where the crest of one wave meets the trough of another wave, they cancel each other and the surface of the water along this line of

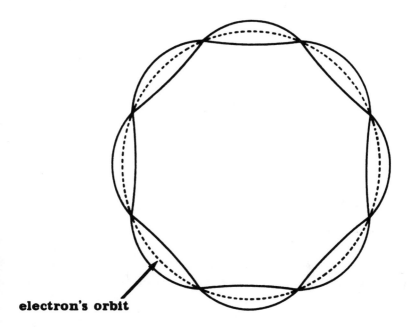

electron's orbit

interaction is calm. These calm areas are the nodes which separate the standing waves. In the double-slit experiment, the nodes are the dark bands in the pattern of alternating light and dark areas. The light bands are the crests of the standing waves.

Schrödinger chose the model of a small tub of water with its complex and intricate interference pattern to explain the nature of the atom. This model is, as he put it, an "analogue" of electron waves in an atom-sized basin.

> The ingenious but nevertheless somewhat artificial assumptions of [Bohr's model of the atom] . . . are replaced by a much more natural assumption in de Broglie's wave phenomena. The wave phenomenon forms the real "body" of the atom. It replaces the individual punctiform [pointlike] electrons, which in Bohr's model swarm around the nucleus.[4]

Standing waves on clotheslines have two dimensions: length and width. Standing waves in mediums like water, or on the head of a conga drum, have three dimensions: length, width, and depth. Schrödinger analyzed the standing wave patterns of the simplest atom, hydrogen, which has only one electron. In hydrogen alone he calculated, using his new wave equation, a multitude of different possible shapes of standing waves. All of the standing waves on a rope are identical. This is not true of the standing waves in an atom. All of them are three-dimensional and all of them are different. Some of them look like concentric circles. Some of them look like butterflies, and others look like mandalas, as in the illustration on the next page.

Shortly before Schrödinger's discovery, another Austrian physicist, Wolfgang Pauli, discovered that no two electrons in an atom can be exactly alike. The presence of an electron with one particular set of properties ("quantum numbers") excludes the presence of another electron with exactly the same properties (quantum numbers) within the same atom. For this reason, Pauli's discovery became known as the Pauli exclusion principle. In terms of Schrödinger's standing wave theory, Pauli's exclusion principle means that once a particular wave pattern forms in an atom, it excludes all others of its kind.

Schrödinger's equation, modified by Pauli's discovery, shows that there are only two possible wave patterns in the lowest of Bohr's energy levels, or shells. Therefore, there can be only two electrons in it. There are eight different standing-wave patterns possible in the next energy level, therefore there can be only eight electrons in it, and so on.

*,†

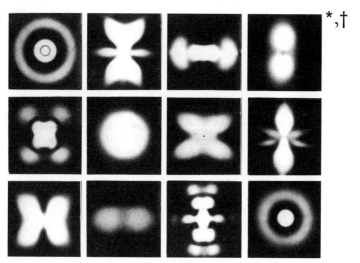

From *Modern College Physics,* Harvey White, N.Y., Van Nostrand, 1972.

These are exactly the numbers of electrons that Bohr's model assigns to these energy levels. In this respect, the two models are alike. In another important way, however, they are different.

Bohr's theory was entirely empirical. That is, he built it around the experimentally observed facts to explain them. In contrast, Schrödinger built his theory on de Broglie's matter-wave hypothesis. Not only does it yield mathematical values which have been verified experimentally, but it also provides a consistent explanation for them.

*These photographs are of mechanical simulations of probability density distributions of different electron states in the hydrogen atom. In other words, they represent where we are most likely to find the point-like electron when we look for it if the atom is in this or that particular state (there are more states than those shown). Initially, Schrödinger pictured electrons as being tenuous clouds actually assuming these patterns.

†A "quantum jump" can be thought of as a transition from one of these pictures to another *without anything in between.*

For example, there are only a certain number of electrons in each energy level, because there are only a certain number of standing wave patterns possible at each energy level. The energy level of an atom jumps only from certain specific values to other certain specific values, because standing-wave patterns of only certain dimensions can form with the atom, and none other.

Although Shrödinger was sure that electrons were standing waves, he was not sure what was waving.* He was convinced, nonetheless, that *something* was waving, and he called it *psi,* a Greek letter pronounced "sigh." (A "wave function" and a "psi function" are the same thing.)

To use the Schrödinger wave equation, we feed it certain characteristics of the atom in question. It then gives us the evolution in time of standing-wave patterns which occur in the atom. If we prepare an atom in an initial state and let it propagate in isolation, that initial state, while propagating in isolation, evolves in time into different standing-wave patterns. The order of these patterns is calculable. The Schrödinger wave equation is the mathematical device which physicists use to calculate the order of these patterns. Said another way, the development of standing wave patterns in an atom is deterministic. Given initial conditions, one pattern always follows another in accordance with the Schrödinger wave equation.†

The Schrödinger wave equation also provides a self-consistent explanation of the size of the hydrogen atom. According to it, the

---

*Schrödinger's early interpretation that electrons literally were standing waves did not stand up to detailed examination and he had to renounce it. Soon, however, the concept of probability based upon a wave function representing an observed system (and developing according to the Schrödinger wave equation) became a fundamental tool in atomic research and Schrödinger's famous equation became an integral part of quantum theory. Since the Schrödinger wave equation is nonrelativistic, however, it does not work at high energies. Therefore, high-energy particle physicists usually use the S Matrix to calculate transition probabilities. (S Matrix theory is discussed in a later chapter.)

†Until the propagating system interacts with a measuring device. That causes an abrupt, unpredictable transition to another state (a quantum jump).

wave pattern of a system with one electron and one proton, which is what we call a hydrogen atom, in its lowest energy state, has an appreciable magnitude only within a sphere which is just the diameter of the smallest Bohr orbit. In other words, such a wave pattern turns out to be the same size as the ground state of a hydrogen atom!

Although Schrödinger's wave mechanics became a pillar of today's quantum mechanics, the useful aspects of Bohr's model of subatomic phenomena still are used when the wave theory does not yield appropriate results. In such cases, physicists simply stop thinking in terms of standing waves and start thinking again in terms of particles. No one can say that they are not adaptable in this matter (wave).

Schrödinger was convinced that his equations described real things, and not mathematical abstractions. He pictured electrons as actually being spread out over their wave patterns in the form of a tenuous cloud. If the picture is limited to the one-electron hydrogen atom, whose standing waves have only three dimensions (length, width, and depth), this is possible to imagine. However, the standing waves in an atom with two electrons exist in six mathematical dimensions; the standing waves in an atom with four electrons exist in twelve dimensions, etc. To visualize this is quite an exercise.

At this point Max Born, a German physicist, put the final touch to the new wave interpretation of subatomic phenomena. According to him, it is not necessary or possible to visualize these waves because they are *not* real things, they are *probability waves.*

> . . . the whole course of events is determined by the laws of probability; to a state in space there corresponds a definite probability, which is given by the de Broglie wave associated with the state.[5]

To obtain the probability of a given state we square (multiply by itself) the amplitude of the matter wave associated with the state.

The question of whether de Broglie's equations and Schrödinger's equations represent real things or abstractions was clear to Born. It did not make sense to him to try to think of a real thing that exists in more than three dimensions.

We have two possibilities. Either we use waves in spaces of more than three dimensions . . . or we remain in three-dimensional space, but give up the simple picture of the wave amplitude as an ordinary physical magnitude, and replace it by a purely abstract mathematical concept . . . into which we cannot enter.[6]

This is exactly what he did. "Physics," he wrote,

is in the nature of the case indeterminate, and therefore the affair of statistics.[7]

This is the same idea (probability waves) that Bohr, Kramers, and Slater had thought of earlier. This time, however, using the mathematics of de Broglie and Schrödinger, the numbers came out right.

Born's contribution to Schrödinger's theory is what enables quantum mechanics to predict probabilities. Since the probability of a state is found by squaring the amplitude of the matter wave associated with it, and, given initial conditions, the Schrödinger equation predicts the evolution of these wave patterns, the two taken together give a determinable evolution of probabilities. Given any initial state, physicists can predict the probability that an observed system will be observed to be in any other given state at any particular time. Whether or not the observed system is observed to be in that state, however, even if that state is the most probable state for that time, is a matter of chance. In other words, the "probability" of quantum mechanics is the probability of observing an observed system in a given state at a given time if it was prepared in a given initial state.*

Thus it was that the wave aspect of quantum mechanics developed. Just as waves have particle-like characteristics (Planck, Einstein), particles also have wave-like characteristics (de Broglie). In fact, particles can be understood in terms of standing waves (Schrödinger).

---

*If the state is prepared in state $\Psi(t)$, the probability that it will be observed to be in state $\phi(t)$ is $| <\Psi(t)|\phi(t)> |^2$. If it is prepared in state $\Psi(t)$ then the probability that it will be observed in region Δ at time t is $\Delta\int_d{}^3 \times \Psi^*(x, t) \times \Psi(x, t)$.

Given initial conditions, a precise evolution of standing-wave patterns can be calculated via the Schrödinger wave equation. Squaring the amplitude of a matter wave (wave function) gives the probability of the state that corresponds to that wave (Born). Therefore, a sequence of probabilities can be calculated from initial conditions by using the Schrödinger wave equation and Born's simple formula.

We have come a long way from Galileo's experiments with falling bodies. Each step along the path has taken us to a higher level of abstraction: first to the creation of things that no one has ever seen (like electrons), and then to the abandonment of all attempts even to picture our abstractions.

The problem is, however, that human nature being what it is, we do not stop trying to picture these abstractions. We keep asking "What are these abstractions of?" and then we try to visualize whatever that is.

Earlier we dismissed Bohr's planetary model of the atom with the promise that we later would see "how physicists currently think of an atom." Well, the time has come, but the task is a thorny one. We gave up our old picture of the atom so easily because we assumed that it would be replaced by one more meaningful, but equally as lucid. Now it develops that our replacement picture is not a picture at all, but an unvisualizable abstraction. This is uncomfortable because it reminds us that atoms were never "real" things anyway. Atoms are hypothetical entities constructed to make experimental observations intelligible. No one, not one person, has ever seen an atom. Yet we are so used to the *idea* that an atom is a thing that we forget that it is an idea. Now we are told that not only is an atom an idea, it is an idea that we cannot even picture.

Nonetheless, when physicists refer to mathematical entities in English (or German or Danish), the words that they use are bound to create images for laymen who hear them, but who are not familiar with the mathematics to which they refer. Therefore, given this lengthy explanation of why it cannot be done, we come now to how physicists today picture an atom.

An atom consists of a nucleus and electrons. The nucleus is located at the center of the atom. It occupies only a small part of the atom's volume, but almost all of its mass. This is the same nucleus as in the planetary model. As in the planetary model, electrons move in the general area of the nucleus. In this model, however, the electrons may be anywhere within an "electron cloud." The electron cloud is made of various standing waves which surround the nucleus. These standing waves are not material. They are patterns of potential. The shape of the various standing waves which comprise the electron cloud tells physicists the probability of finding the point electron at any given place in the cloud.

In short, physicists still think of an atom as a nucleus around which move electrons, but the picture is not so simple as that of a tiny solar system. The electron cloud is a mathematical concept which physicists have constructed to correlate their experiences. Electron clouds may or may not exist within an atom. No one really knows. However, we do know that the concept of an electron cloud yields the probabilities of finding the electron at various places around the nucleus of an atom, and that these probabilities have been determined empirically to be accurate.

In this sense, electron clouds are like wave functions. A wave function also is a mathematical concept which physicists have constructed to correlate their experiences. Wave functions may or may not "actually exist." (This type of statement assumes a qualitative difference between thought and matter, which may not be a good assumption). However, the concept of a wave function undeniably yields the probabilities of observing a system to be in a given state at a given time if it was prepared in a given way.

Like wave functions, electron clouds generally cannot be visualized. An electron cloud containing only one electron (like the electron cloud of a hydrogen atom) exists in three dimensions. All other electron clouds, however, contain more than one electron and therefore exist in more than three dimensions. The nucleus of the simple carbon atom, for example, with its six electrons, is surrounded by an electron cloud with eighteen dimensions. Uranium, with ninety-two electrons, has an electron cloud of two hundred and seventy-six dimensions. (Similarly, a wave function contains three dimensions for each particle that it represents). The situation, in terms of mental pictures, is clearly unclear.

This ambiguity results from attempting to depict with limited concepts (language) situations which are not bound by the same limitations. It also masks the fact that *we do not know* what actually is going on in the invisible subatomic realm. The models that we use are "free creations of the human mind," to use Einstein's words (page 9), that satisfy our innate need to correlate experience rationally. They are guesses about what "really" goes on inside the unopenable watch. It is extremely misleading to think that they actually describe anything.

In fact, a young German physicist, Werner Heisenberg, decided that we *never* can know what actually goes on in the invisible subatomic realm, and that, therefore, we should "abandon all attempts to construct perceptual models of atomic processes."[8] All that we legitimately can work with, according to this theory, is what we observe directly. All we know is what we have at the beginning of an experiment, and what we have at the end of it. Any explanation of what actually happens between these two states—which are the observables (page 77)—is speculation.

Therefore, about the same time (1925), but independently of de Broglie and Schrödinger, the twenty-five-year-old Heisenberg set about developing a means of organizing experimental data into tabular form. He was fortunate in that sixty-six years earlier an Irish mathematician named W. R. Hamilton had developed a method of organizing data into arrays, or mathematical tables, called matrices. At that time, Hamilton's matrices were considered the fringe of pure mathematics. Who could have guessed that one day they would fit, like a precut piece, into the structure of a revolutionary physics?

To use Heisenberg's tables, we simply read from them, or calculate from them, what probabilities are associated with what initial conditions. Using this method, which Heisenberg called matrix mechanics, we deal only with physical observables, which means those things that we know at the beginning of an experiment, and those things that we know about it at the end. We make no speculation about what happens in between.

After twenty-five years of struggling for a theory to replace Newtonian physics, physicists suddenly found themselves with two different theories, each one a unique way of approaching the same thing: Schrödinger's wave mechanics, based on de Broglie's matter waves, and Heisenberg's matrix mechanics, based on the unanalyzability of subatomic phenomena.

Within a year after Heisenberg developed his matrix mechanics, Schrödinger discovered that it was mathematically equivalent to his own wave mechanics. Since both of these theories were valuable tools for subatomic research, both of them were incorporated into the new branch of physics which became known as quantum mechanics.

Much later, Heisenberg applied matrix mathematics to the particle collision experiments of high-energy particle physics. Because such collisions always result in a scattering of particles, it was called the Scattering Matrix, which was shortened to the S Matrix. Today, physicists have two ways to calculate the transition probabilities between what they observe at the beginning of a quantum mechanical experiment and what they observe at the end of it.

The first method is the Schrödinger wave equation, and the second method is the S Matrix. The Schrödinger wave equation describes a temporal development of possibilities, one of which suddenly actualizes when we make a measurement in the course of a quantum mechanical experiment. The S Matrix gives directly the transition probabilities between the observables without giving any indication of a development in time, or the lack of it, or anything else. Both of them work.*

As important as was Heisenberg's introduction of matrix mathematics into the new physics, his next discovery shook the very founda-

---

*The Schrödinger wave equation works at lower energies, however, since it is nonrelativistic, it does not work for high energies. Therefore, most particle physicists use the S Matrix together, perhaps, with local relativistic quantum fields to understand quarks and particles.

tions of "the exact sciences." He proved that, at the subatomic level, there is no such thing as "the exact sciences."

Heisenberg's remarkable discovery was that there are limits beyond which we cannot measure accurately, at the same time, the processes of nature. These limits are not imposed by the clumsy nature of our measuring devices or the extremely small size of the entities that we attempt to measure, but rather by the very way that nature presents itself to us. In other words, there exists an ambiguity barrier beyond which we never can pass without venturing into the realm of uncertainty. For this reason, Heisenberg's discovery became known as the "uncertainty principle."

The uncertainty principle reveals that as we penetrate deeper and deeper into the subatomic realm, we reach a certain point at which one part or another of our picture of nature becomes blurred, and there is no way to reclarify that part without blurring another part of the picture! It is as though we are adjusting a moving picture that is slightly out of focus. As we make the final adjustments, we are astonished to discover that when the right side of the picture clears, the left side of the picture becomes completely unfocused and nothing in it is recognizable. When we try to focus the left side of the picture, the right side starts to blur and soon the situation is reversed. If we try to strike a balance between these two extremes, both sides of the picture return to a recognizable condition, but in no way can we remove the original fuzziness from them.

The right side of the picture, in the original formulation of the uncertainty principle, corresponds to the position in space of a moving particle. The left side of the picture corresponds to its momentum. According to the uncertainty principle, we cannot measure accurately, at the same time, both the position *and* the momentum of a moving particle. The more precisely we determine one of these properties, the less we know about the other. If we precisely determine the position of the particle, then, strange as it sounds, there is *nothing* that we can know about its momentum. If we precisely determine the momentum of the particle, there is no way to determine its position.

To illustrate this strange statement, Heisenberg proposed that we imagine a super microscope of extraordinarily high resolving power—powerful enough, in fact, to be able to see an electron moving around in its orbit. Since electrons are so small, we cannot use ordinary light in our microscope because the wavelength of ordinary light is much too long to "see" electrons, in the same way that long sea waves barely are influenced by a thin pole sticking out of the water.

If we hold a strand of hair between a bright light and the wall, the hair casts no distinct shadow. It is so thin compared to the wavelengths of the light that the light waves bend around it instead of being obstructed by it. To see something, we have to obstruct the light waves we are looking with. In other words, to see something, we have to illuminate it with wavelengths smaller than it is. For this reason, Heisenberg substituted gamma rays for visible light in his imaginary microscope. Gamma rays have the shortest wavelength known, which is just what we need for seeing an electron. An electron is large enough, compared to the tiny wavelength of gamma rays, to obstruct some of them: to make a shadow on the wall, as it were. This enables us to locate the electron.

The only problem, and this is where quantum physics enters the picture, is that, according to Planck's discovery, gamma rays, which have a much shorter wavelength than visible light, also contain much more energy than visible light. When a gamma ray strikes the imaginary electron, it illuminates the electron, but unfortunately, it also knocks it out of its orbit and changes its direction and speed (its momentum) in an unpredictable and uncontrollable way. (We cannot calculate precisely the angle of rebound between a particle, like the electron, and a wave, like the gamma ray). In short, if we use light with a wavelength short enough to locate the electron, we cause an undeterminable change in the electron's momentum.

The only alternative is to use a less energetic light. Less energetic light, however, causes our original problem: Light with an energy low enough not to disturb the momentum of the electron will have a wavelength so long that it will not be able to show us where the electron is! There is no way that we can know simultaneously the position *and* the momentum of a moving particle. All attempts to observe the electron alter the electron.

This is the primary significance of the uncertainty principle. At the subatomic level, *we cannot observe something without changing it.* There is no such thing as the independent observer who can stand on the sidelines watching nature run its course without influencing it.

In one sense, this is not such a surprising statement. A good way to make a stranger turn and look at you is to stare intently at his back. All of us know this, but we often discredit what we know when it contradicts what we have been taught is possible. Classical physics is based on the assumption that our reality, independently of us, runs its course in space and time according to strict causal laws. Not only can we observe it, unnoticed, as it unfolds, we can predict its future by applying causal laws to initial conditions. In this sense, Heisenberg's uncertainty principle is a *very* surprising statement.

We cannot apply Newton's laws of motion to an individual particle that does not have an initial location and momentum, which is exactly what the uncertainty principle shows us that we cannot determine. In other words, it is impossible, even in principle, ever to know enough about a particle in the subatomic realm to apply Newton's laws of motion which, for three centuries, were the basis of physics. *Newton's laws do not apply to the subatomic realm.** (Newton's *concepts* do not even apply in the subatomic realm.) Given a beam of electrons, quantum theory can predict the probable distribution of the electrons over a given space at a given time, but quantum theory cannot predict, even in principle, the course of a single electron. The whole idea of a causal universe is undermined by the uncertainty principle.

In a related context, Niels Bohr wrote that quantum mechanics, by its essence, entails:

---

*Strictly speaking, Newton's laws do not disappear totally in the subatomic realm: they remain valid as operator equations. Also, in some experiments involving subatomic particles Newton's laws may be taken as good approximations in the description of what is happening.

. . . the necessity of a final renunciation of the classical ideal of causality and a radical revision of our attitude toward the problem of physical reality.[9]

Yet there is another startling implication in the uncertainty principle. The concepts of position and momentum are intimately bound up with our idea of a thing called a moving particle. If, as it turns out, we cannot determine the position and momentum of a moving particle, as we always have assumed that we could, then we are forced to admit that this thing that we have been calling a moving particle, whatever it is, is *not* the "moving particle" we thought it was, because "moving particles" always have both position and momentum.

As Max Born put it:

. . . if we can never actually determine more than one of the two properties (possession of a definite position and of a definite momentum), and if when one is determined we can make no assertion at all about the other property for the same moment, so far as our experiment goes, then we are not justified in concluding that the "thing" under examination can actually be described as a particle in the usual sense of the term.[10]

Whatever it is that we are observing *can* have a determinable momentum, and it *can* have a determinable position, but of these two properties, *we must choose,* for any given moment, which one we wish to bring into focus. This means, in reference to "moving particles" anyway, that we can never see them the way they "really are," but only the way we choose to see them!

As Heisenberg wrote:

What we observe is not nature itself, but nature exposed to our method of questioning.[11]

The uncertainty principle rigorously brings us to the realization that there is no "My Way" which is separate from the world around us. It brings into question the very existence of an "objective" reality, as does complementarity and the concept of particles as correlations.

The tables have been turned. "The exact sciences" no longer study an objective reality that runs its course regardless of our interest in it or not, leaving us to fare as best we can while it goes its predetermined way. Science, at the level of subatomic events, is no longer "exact," the distinction between objective and subjective has vanished, and the portals through which the universe manifests itself are, as we once knew a long time ago, those impotent, passive witnesses to its unfolding, the "I"s, of which we, insignificant we, are examples. The Cogs in the Machine have become the Creators of the Universe.

If the new physics has led us anywhere, it is back to ourselves, which, of course, is the only place that we could go.

Part One

NONSENSE

1

🌀

Beginner's Mind

The importance of nonsense hardly can be overstated. The more clearly we experience something as "nonsense," the more clearly we are experiencing the boundaries of our own self-imposed cognitive structures. "Nonsense" is that which does not fit into the prearranged patterns which we have superimposed on reality. There is no such thing as "nonsense" apart from a judgmental intellect which calls it that.

True artists and true physicists know that nonsense is only that which, viewed from our present point of view, is unintelligible. Nonsense is nonsense only when we have not yet found that point of view from which it makes sense.

In general, physicists do not deal in nonsense. Most of them spend their professional lives thinking along well-established lines of thought. Those scientists who establish the established lines of thought, however, are those who do not fear to venture boldly into nonsense, into that which any fool could have told them is clearly not so. This is the mark of the creative mind; in fact, this *is* the creative process. It is characterized by a steadfast confidence that there exists a point of view from which the "nonsense" is not nonsense at all—in fact, from which it is obvious.

In physics, as elsewhere, those who most have felt the exhilaration of the creative process are those who best have slipped the bonds of the known to venture far into the unexplored territory which lies beyond the barrier of the obvious. This type of person has two characteristics. The first is a childlike ability to see the world as it is, and not as it appears according to what we know about it. This is the moral of the (child's?) tale, "The Emperor's New Clothes." When the emperor rode naked through the streets, only a child proclaimed him to be without clothes, while the rest of his subjects forced themselves to believe, because they had been told so, that he wore his finest new clothing.

The child in us is always naive, innocent in the simplistic sense. A Zen story tells of Nan-in, a Japanese master during the Meiji era who received a university professor. The professor came to inquire about Zen. Nan-in served tea. He poured his visitor's cup full, and then kept on pouring. The professor watched the overflow until he no longer could restrain himself.

"It is overfull. No more will go in!"

"Like this cup," Nan-in said, "you are full of your own opinions and speculations. How can I show you Zen unless you first empty your cup?"

Our cup usually is filled to the brim with "the obvious," "common sense," and "the self-evident."

Suzuki Roshi, who established the first Zen center in the United States (without trying, of course, which is very Zen), told his students that it is not difficult to attain enlightenment, but it is difficult to keep a beginner's mind. "In the beginner's mind," he told them, "there are any possibilities, but in the expert's there are few." When his students published Suzuki's talks after his death, they called the book, appropriately, *Zen Mind, Beginner's Mind*. In the introduction, Baker Roshi, the American Zen Master, wrote:

> The mind of the beginner is empty, free of the habits of the expert, ready to accept, to doubt, and open to all the possibilities. . . .[1]

The beginner's mind in science is wonderfully illustrated by the story of Albert Einstein and his theory of relativity. That is the subject of this chapter.

The second characteristic of true artists and true scientists is the firm confidence which both of them have in themselves. This confidence is an expression of an inner strength which allows them to speak out, secure in the knowledge that, appearances to the contrary, it is the world that is confused and not they. The first man to see an illusion by which men have flourished for centuries surely stands in a lonely place. In that moment of insight he, and he alone, sees the obvious which to the uninitiated (the rest of the world) yet appears as nonsense or, worse, as madness or heresy. This confidence is not the obstinacy of the fool, but the surety of him who knows what he knows, and knows also that he can convey it to others in a meaningful way.

The writer, Henry Miller, wrote:

I obey only my own instincts and intuition. I know nothing in advance. Often I put down things which I do not understand myself, secure in the knowledge that later they will become clear and meaningful to me. I have faith in the man who is writing, who is myself, the writer.[2]

The songwriter Bob Dylan told a press conference:

I just write a song and I know it's going to be all right. I don't even know what it's going to say.[3]

An example of this kind of faith in the realm of physics was the theory of light quanta. In 1905, the accepted and proven theory of light was that light was a wave phenomenon. In spite of this, Einstein published his famous paper proposing that light was a particle phenomenon (page 58). Heisenberg described this fascinating situation this way:

[In 1905] light could either be interpreted as consisting of electromagnetic waves, according to Maxwell's theory, or as consisting of light quanta, energy packets traveling through space with high velocity [according to Einstein]. But could it be both? Einstein knew, of course, that the well-known phenomena of diffraction and interference can be explained only on the basis of the wave picture. He was not able to dispute the complete contradiction between this wave picture and the idea of the light quanta; nor did he even attempt to remove the inconsistency of this interpretation. He simply took the contradiction as something which would probably be understood much later.[4]

That is exactly what happened. Einstein's thesis led to the wave-particle duality from which quantum mechanics emerged, and with it, as we know, a way of looking at reality and ourselves that is vastly different from that to which we were accustomed. Although Einstein is known popularly for his theories of relativity, it was his paper on the quantum nature of light that won him the Nobel Prize. It is also a fine example of confidence in nonsense.

What is nonsense and what is not, then, may be merely a matter of perspective.

"Wait a minute," interrupts Jim de Wit. "My uncle, Weird George, believes that he is a football. Of course, we know that this is nonsense, but Uncle George thinks that *we* are mad. He is quite certain that he is a football. He talks about it constantly. In other words, he has abundant confidence in his nonsense. Does this make him a great scientist?

No. In fact, Weird George has a problem. Not only is he the only person who has this particular perspective, but also this particular perspective is in no way relative to that of any other observer, which brings us to the heart of Einstein's special theory of relativity. (Einstein created two theories of relativity. The first theory is called the special theory of relativity. The second theory, which came later and is more general, is called the general theory of relativity. This chapter and the next are about the first theory, the special theory of relativity).

The special theory of relativity is not so much about what is relative as about what is not. It describes in what way the relative aspects of physical reality appear to vary, depending upon the point of view of different observers (actually depending upon their state of motion relative to each other), but, in the process, it defines the nonchanging, absolute aspect of physical reality as well.

The special theory of relativity is not a theory that everything is relative. It is a theory that *appearances* are relative. What may appear to us as a ruler (physicists say "rod") one foot long, may appear to an observer traveling past us (very fast) as being only ten inches long. What may appear to us as one hour, may appear to an observer traveling past us (very fast) as two hours. However, the moving observer can use the special theory of relativity to determine how our ruler and our clock appear to us (if he knows his motion relative to us) and, likewise, we can use the special theory of relativity to determine how our stick and our clock appear to the moving observer (if we know our motion relative to him).

If we were to perform an experiment at the same moment that the moving observer came past us, both we and the moving observer would see the same experiment, but each of us would record different times and distances, we with our rod and clock and he with his rod and clock. Using the special theory of relativity, however, each of us could transpose our data to the other's frame of reference. The final numbers would come out the same for both of us. In essence, the special theory of relativity is not about what is relative, it is about what is absolute.

However, the special theory of relativity does show that appearances are dependent upon the state of motion of the observers. For example, the special theory of relativity tells us that (1) a moving object measures shorter in its direction of motion as its velocity increases until, at the speed of light, it disappears; (2) the mass of a moving object measures more as its velocity increases until, at the speed of light, it becomes infinite; and (3) moving clocks run more slowly as their velocity increases until, at the speed of light, they stop running altogether.

All of this is from the point of view of an observer to whom the object is moving. To an observer traveling along with the moving object, the clock keeps perfect time, ticking off sixty seconds each minute, and nothing appears to get any shorter or more massive. The special theory of relativity also tells us that space and time are not two separate things, but that together they form space-time, and that energy and mass are actually different forms of the same thing, mass-energy.

"This is not possible!" we cry. "It is nonsense to think that increasing the velocity of an object increases its mass, decreases its length, and slows its time."

Our cup runneth over.

These phenomena are not observable in everyday life because the velocities required to make them noticeable are those approaching the speed of light (186,000 miles per *second*). At the slow speeds that we encounter in the macroscopic world, these effects are virtually undetectable. If they were, we would discover that a car traveling down the freeway is shorter than it is at rest, weighs more than it does at rest, and that its clock runs slower than it does at rest. In fact, we even would find that a hot iron weighs more than a cold one (because energy has mass and heat is energy).

How Einstein discovered all of this is another version of "The Emperor's New Clothes."

Only Albert Einstein looked at two of the major puzzles of his day and saw them with a beginner's mind. The result was the special theory of relativity. The first puzzle of Einstein's time was the constancy of the speed of light. The second puzzle of Einstein's time was the uncertainty, both physical and philosophical, about what it means to be moving or not moving.*

*Einstein's point of departure for the special theory of relativity came from the conflict of classical relativity and Maxwell's prediction of a light speed, "c." An often-told story tells how Einstein tried to imagine what it would be like to travel as fast as a light wave. He saw, for example, that the hands on a clock would

"Wait a minute," we say. "What is uncertain about that? If I am sitting in a chair and another person walks past me, then the person walking past me is in motion, and I, sitting in my chair, am not in motion."

"Quite right," says Jim de Wit, appearing on cue, "but still, it is not that simple. Suppose that the chair in which you are sitting is on an airplane and that the person walking past you is a stewardess. Suppose also that I am on the ground watching both of you go by. From your point of view, you are at rest and the stewardess is in motion, but from my point of view, I am at rest and *both* of you are in motion. It all depends upon your frame of reference. Your frame of reference is the airplane, but my frame of reference is the earth."

De Wit, as usual, has discovered the problem exactly. Unfortunately, he has not solved it. The earth itself hardly is standing still. Not only is it spinning on its axis like a top, it and the moon are revolving around a common center of gravity while both of them circle the sun at eighteen miles per second.

"That's not fair," we say. "Of course, it is true, but the earth does not seem to be moving to us who live on it. It is only in motion if we change our frame of reference from it to the sun. If we start playing that game, it is impossible to find anything in the entire universe that is 'standing still.' From the point of view of the galaxy, the sun is moving; from the point of view of another galaxy, our galaxy is moving; from the point of view of a third galaxy, the first two galaxies are moving. In fact, from the point of view of each of them, the others are moving."

"Nicely said," laughs Jim de Wit, "and that is exactly the point. There is no such thing as something being absolutely at rest, unequivocally not moving. Motion, and the lack of it, is always relative to something else. Whether we are moving or not depends upon what frame of reference we use."

The discussion above is *not* the special theory of relativity. In fact,

---

appear to stand still, since no other light waves from the clock would be able to catch up with him until he slowed down.

the discussion above is a part of the Galilean relativity principle which is over three hundred years old. Any physical theory is a theory of relativity if, like Jim de Wit, it acknowledges the difficulty of detecting absolute motion or absolute nonmotion. A theory of relativity assumes that the only kind of motion that we ever can determine is motion, or lack of it, relative to something else. Galileo's principle of relativity says, in addition, that the laws of mechanics are equally valid in all frames of reference (physicists say "co-ordinate systems") that move uniformly in relation to each other.

The Galilean relativity principle assumes that somewhere in the universe there exists a frame of reference in which the laws of mechanics are completely valid—that is, a frame of reference in which experiment and theory agree perfectly. This frame of reference is called an "inertial" frame of reference. An inertial frame of reference simply means a frame of reference in which the laws of mechanics are completely valid. All other frames of reference moving uniformly, relative to an inertial frame of reference, are also inertial frames of reference. Since the laws of mechanics are equally valid in all inertial frames of reference, this means that there is no way that we can distinguish between one inertial frame of reference and another by performing mechanical experiments in them.

Frames of reference moving uniformly, relative to each other, are co-ordinate systems that move with a constant speed and direction. In other words, they are frames of reference that move with a constant velocity. For example, if, by accident, we drop a book while standing in line at the library, the book falls directly downward in accordance with Newton's law of gravity, and strikes the ground directly beneath the place from which it was dropped. Our frame of reference is the earth. The earth is moving at a fantastic speed on its trip around the sun, but this speed is constant.*

If we drop the same book while we are traveling on an ideally smooth train which is moving at a constant speed, the same thing

---

*Although we do not experience it directly, the orbital motion of the earth is accelerating.

happens. The book falls directly downward in accordance with Newton's laws of gravitation, and strikes the floor of the train directly beneath the place from which it was dropped. This time, our frame of reference is the train. Because the train is moving uniformly, with no increases or decreases of speed, in relation to the earth, and because the earth is moving in a similar manner in relation to the train, the two frames of reference are moving uniformly relative to each other, and the laws of mechanics are valid in both of them. It does not matter in the least which of the frames of reference is "moving." A person in either frame of reference can consider himself moving and the other frame of reference at rest (the earth is at rest and the train is moving) or the other way round (the train is at rest and the earth is moving). From the point of view of physics, there is no difference.

What happens if the engineer suddenly accelerates while we are doing our experiment? Then, of course, everything is upset. The falling book still will strike the floor of the train, but at a spot farther back since the floor of the train has moved forward beneath the book while it was falling. In this case, the train is not moving uniformly in relation to the earth, and the Galilean relativity principle does not apply.

Provided that all of the motion involved is uniformly relative, we can translate motion as perceived in one frame of reference into another frame of reference. For example, suppose that we are standing on the shore watching a ship move past us at thirty miles per hour. The ship is a frame of reference moving uniformly relative to us. There is a passenger, a man, standing on the deck of the ship, leaning against the railing. Since he is standing still, his velocity is the same as that of the ship, thirty miles per hour. (From his point of view, we are moving past *him* at thirty miles per hour).

Suppose now that the man begins to walk toward the front of the ship at three miles per hour. His velocity now, relative to us, is thirty-three miles per hour. The ship carries him forward at thirty miles per hour, and his walking adds three miles per hour to that. (You get to the top of an escalator faster if you walk.)

Suppose that the man turns around and walks back toward the

rear of the ship. His velocity relative to the ship is, again, three miles per hour, but his velocity relative to the shore is now twenty-seven miles per hour.

In other words, to calculate how fast this passenger moves relative to us, we add his velocity to the velocity of his co-ordinate system (the ship) if he is walking in the same direction that it is moving, and we subtract his velocity from the velocity of his co-ordinate system if he is walking in the opposite direction. This calculation is called a classical (Galilean) transformation. Knowing the uniform relative motion of our two frames of reference, we can transform the passenger's velocity in reference to his own co-ordinate system (three miles per hour) into his velocity in reference to our co-ordinate system (thirty-three miles per hour).

The freeway provides abundant examples of classical transformations from one frame of reference to another. Suppose that we are driving at 75 miles per hour. We see a truck coming toward us. Its speedometer also reads 75 miles per hour. Making a classical transformation, we can say that, relative to us, the truck is approaching at 150 miles per hour, which explains why head-on collisions so often are fatal.

Suppose now that a car going in the same direction that we are going passes us. His speedometer reads 110 miles per hour (it's a Ferrari). Again, making a classical transformation, we can say that, relative to us, the Ferrari is departing our location at 35 miles per hour.

The transformation laws of classical mechanics are common sense. They say that, even though we cannot determine whether a frame of reference is absolutely at rest or not, we can translate velocities (and positions) from one frame of reference into velocities (and positions) in other frames of reference, provided that the frames of reference are moving uniformly, relative to each other. Furthermore, the Galilean relativity principle, from which Galilean transformations come, says that if the laws of mechanics are valid in any one frame of reference, they also are valid in any other frame of reference moving uniformly relative to it.

Unfortunately, there is one catch in all this. No one yet has found a co-ordinate system in which the laws of mechanics are valid!*

"What! Impossible! Can't be!" we cry, aghast. "What about the earth?"

Well, it is true that Galileo, who first probed the laws of classical mechanics, used the earth as a frame of reference, although not consciously. (The idea of co-ordinate systems did not come along until Descartes). However, our present measuring devices are more accurate than Galileo's, who occasionally even used his pulse (which means that the more excited he got, the more inaccurate his measurements became!). Whenever we reconstruct Galileo's falling body experiments, we always find discrepancies between the theoretical results that we should get and the experimental results that we actually do get. These discrepancies are due to the rotation of the earth. The bitter truth is that the laws of mechanics are not valid for a co-ordinate system rigidly attached to the earth. The earth is *not* an inertial frame of reference. Since their very inception, the poor laws of classical mechanics have been left, so to speak, without a home. No one has discovered a co-ordinate system in which they manifest themselves perfectly.

This leaves us, from a physicist's point of view, in a pretty mess. On the one hand, we have the laws of classical mechanics, which are indispensable to physics, and, on the other hand, these same laws are predicated upon a co-ordinate system which may not even exist.

This problem is related to relativity, which is the problem of determining absolute nonmotion, in an intimate way. If such a thing as absolute nonmotion were detected, then a co-ordinate system attached to it would be the long-lost inertial frame of reference, the co-ordinate system in which the classical laws of mechanics are perfectly valid. Then everything would make sense again because, given a frame of reference in which the classical laws of mechanics are valid, *any* frame of reference, the classical laws of mechanics at last would have a permanent mailing address.

*The fixed stars provide such a reference frame as far as defining nonrotation.

Physicists do not enjoy theories with loose ends. Before Einstein, the problem of detecting absolute motion (or absolute non-motion—if we find one, we find the other), and the problem of finding an inertial co-ordinate system were, to say the least, loose ends. The entire structure of classical mechanics was based on the fact that somewhere, somehow, there must be a frame of reference in which the laws of classical mechanics are valid. The inability of physicists to find it made classical mechanics appear exactly like a huge castle built on sand.

Although no one, including Einstein, discovered absolute non-motion, the inability to detect it was a major concern of Einstein's day. The second major controversy of Einstein's day (not counting Planck's discovery of the quantum) was an incomprehensible, logic-defying characteristic of light.

In the course of their experiments with the speed of light, physicists discovered something very strange. The speed of light disregards the transformation laws of classical mechanics. Of course, that's impossible, but nevertheless, experiment after experiment proved just the opposite. The speed of light just happens to be the most nonsensical thing ever discovered. That is because it never changes.

"So light always travels at the same speed," we ask, "what's so strange about that?"

"Oh my, oh my," says a distraught physicist, circa 1887, "you simply don't understand the problem. The problem is that no matter what the circumstances of the measurement, no matter what the motion of the observer, the speed of light *always* measures 186,000 miles per second."*

"Is this bad?" we say, beginning to sense that something *is* strange here.

---

*In a vacuum. The speed of light changes in matter depending upon the index of refraction of the matter: $c_{matter} = \dfrac{c}{\text{index of refraction}}$

"Worse," says the physicist. "It's impossible. Look," he tells us, trying to calm himself, "suppose that we are standing still and that somewhere in front of us is a light bulb that also is standing still. The light bulb flashes on and off and we measure the velocity of the light that comes from it. What do you suppose that velocity will be?"

"186,000 miles per second," we answer, "the speed of light."

"Correct!" says the physicist, with a knowing look that makes us uncomfortable. "Now, suppose the light bulb still is standing still, but we are moving toward it at 100,000 miles per second. Now what will we measure the speed of the light to be?"

"286,000 miles per second," we answer, "the speed of light (186,000 miles per second) plus our speed (100,000 miles per second)." (This is a typical example of a classical transformation.)

"Wrong!" shouts the physicist. "That's just the point. *The speed of the light is still 186,000 miles per second.*"

"Wait a minute," we say. "That can't be. You say that if the light bulb is at rest and we are at rest, the speed of photons emitted from it will measure the same to us as the speed of photons emitted from it when we are rushing toward the light bulb? That doesn't make sense. When the photons are emitted, they are traveling at 186,000 miles per second. If we also are moving, and moving toward them, their velocity should measure that much faster. In fact, they should appear to be traveling with the speed at which they were emitted *plus* our speed. Their velocity should measure 186,000 miles per second plus 100,000 miles per second."

"True," says our friend, "but it doesn't. It measures 186,000 miles per second, just as if we still were standing still."

Pausing for that to sink in, he continues, "Now consider the opposite situation. Suppose that the light bulb still is standing still, and this time we are moving *away* from it at 100,000 miles per second. What will the velocity of the photons measure now?"

"86,000 miles per second?" we say, hopefully, "the speed of light minus our speed as we move away from the approaching photons?"

"Wrong, again!" exclaims our friend again. "It should, but it doesn't. The speed of the photons still measures 186,000 miles per second."

"This is very hard to believe. Do you mean that if a light bulb is at rest and we measure the speed of the photons emitted from it while we also are at rest, and if we then measure the speed of the photons from it while we are moving toward it, and lastly, if we measure the speed of the photons emitted from it while we are moving away from it, we get *the same result* in all three cases?"

"Exactly!" says the physicist. "186,000 miles per second."*

"Do you have any evidence?" we ask him.

"Unfortunately," he says, "I do. Two American physicists, Albert Michelson and Edward Morley, have just completed an experiment which seems to show that the speed of light is constant, regardless of the state of motion of the observer.

"This can't happen," he sighs, "but it *is* happening. It just doesn't make sense."

The problem of absolute nonmotion and the problem of the constancy of the speed of light converged in the Michelson-Morley experiment. The Michelson-Morley experiment (1887) was a crucial experiment. A crucial experiment is an experiment which determines

---

*The reverse situation (the source moves and the observer remains stationary) is explainable in terms of prerelativistic physics. In fact, if light is assumed to be a wave phenomenon governed by a wave equation, it is *expected* that its measured velocity will be independent of the velocity of its source. The velocity of the sound waves reaching us from a jet plane, for example, does not depend upon the velocity of the aircraft. They propagate through a medium (the atmosphere) at a given velocity, from their point of origin, regardless of the motion of the plane (the *frequency* of the sound shifts as the source moves, e.g., the Doppler effect). Prerelativity theory assumes a medium (like the atmosphere, for sound waves, or the ether, for light waves) through which the waves propagate. The paradox is that the measured velocity of light has been found (the Michelson-Morley experiment) to be independent of the motion of the observer. In other words, assuming a light wave propagating through a medium, how can we move through the same medium toward the approaching wave without increasing its measured velocity?

the life or death of a scientific theory. The theory that was tested by the Michelson-Morley experiment was the theory of the ether.

The theory of the ether was that the entire universe lies in and is permeated by an invisible, tasteless, odorless substance that has no properties at all, and exists simply because it has to exist so that light waves can have something to propagate in. For light to travel as waves, according to the theory, something has to be waving. That something was the ether. The theory of the ether was the last attempt to explain the universe by explaining some*thing*. Interpreting the universe in terms of things (like the Great Machine idea) was the distinguishing characteristic of the mechanical view, which means all of physics from Newton until the middle 1800s.

The ether, according to the theory, is everywhere and in everything. We live and perform our experiments in a sea of ether. To the ether, the hardest substance is as porous as a sponge to water. There are no doors to the ether. Although we move in the ether sea, the ether sea does not move. It is absolutely, unequivocally not moving.

Therefore, although the primary reason for the existence of the ether was to give light something to propagate through, its existence also solved the old problem of locating the original inertial co-ordinate system, that frame of reference in which the laws of mechanics are completely valid. If the ether existed (and it *had* to exist), the co-ordinate system attached to it was *the* co-ordinate system against which all others could be compared to see if they were moving or not.

The findings of Michelson and Morley gave a verdict of death to the theory of the ether.* Equally important, they led to the mathematical foundations of Einstein's revolutionary new theory.

The idea of the Michelson-Morley experiment was to determine the motion of the earth through the ether sea. The problem was how to do this. Two ships at sea can determine their motion relative to one another, but if only one ship moves through a smooth sea, it has no

---

*Quantum field theory resurrects a new kind of ether, e.g., particles are excited states of the featureless ground state of the field (the vacuum state). The vacuum state is so featureless and has such high symmetry that we cannot assign a velocity to it experimentally.

reference point against which to measure its progress. In the old days, seamen would throw a log overboard and measure their progress relative to it. Michelson and Morley did the same thing, except that the log that they threw overboard was a beam of light.

Their experiment was conceptually simple and ingenious. If the earth is moving, they reasoned, and the ether sea is at rest, then the movement of the earth through the ether sea must cause an ether breeze. Therefore, a beam of light traveling against the ether breeze should have a slower velocity than a beam of light sent across the ether breeze. This is the essence of the Michelson-Morley experiment.

Every pilot knows that it takes longer to fly a given distance if one leg of the trip is against a head wind (even though the return leg is with a tail wind) than it takes to fly the same distance across the same wind. Similarly, thought Michelson and Morley, if the theory of the ether sea is correct, a light beam sent upstream against the ether breeze and then downstream with it will take longer to return to its starting point than a light beam sent back and forth across the either breeze.

To establish and detect this difference in velocity, Michelson and Morley created a device called an interferometer (from the word "interference"). It was designed to detect the interference pattern created by the two beams of light as they returned to a common point. (A diagram is on the next page.)

A light source emits a beam of light toward a half-silvered mirror (similar to the lenses in sunglasses that look like a mirror on one side, but are transparent from the other side). The original beam of light (=====▶) is split by the half-silvered mirror into two segments (——▶) (– – – –▶), each of which travels an equal distance, but at right angles to each other, and back again. The two beams then reunite via the same half-silvered mirror and travel (=====▶) into a measuring device. By observing the interference created by these converging beams in the measuring device, any difference in velocity between them can be determined accurately.

When the experiment was performed, not the slightest difference in velocity could be detected between the two beams of light. The

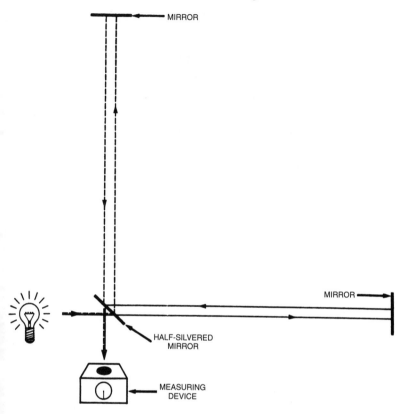

interferometer was turned 90 degrees so that the beam going against the ether wind now was directed across it, and the beam going across the ether wind now was sent directly into it. Again not the slightest difference in velocity between the two beams could be detected.

In other words, the Michelson-Morley experiment had failed to prove the existence of the ether. Unless an explanation could be found, physicists would be faced with choosing between two unsettling alternatives: either (1) the earth is not moving (and Copernicus was wrong), or (2) the ether does not exist. Neither of these was very acceptable.

Michelson and Morley thought that perhaps the earth carried a layer of ether with it as it moved through the ether set, just as it carries its atmosphere with it as it travels through space and, therefore, close to the surface of the earth, the ether breeze cannot be detected. No one had a better hypothesis until an Irishman named George Francis FitzGerald proposed (in 1892) an outrageous explanation.

FitzGerald reasoned that perhaps the pressure of the ether wind compresses matter just as an elastic object moving through water becomes shortened in the direction that it is traveling. If this were true, then the arm of the interferometer pointing into the ether wind would be somewhat shorter than the arm that is not pointing into it. Therefore, a reduction in the velocity of the light traveling into the ether wind and back might not be detected because the distance that the light travels also is reduced. In fact, if the amount by which the interferometer arm pointing into the ether wind is shortened just corresponds to the amount by which the velocity of the light traveling up that arm and back is reduced, then both beams of light in the experiment will reach the measuring device at exactly the same time (the beam with the higher velocity traversing a greater distance in the same time that the beam with the slower velocity traverses a lesser distance).

FitzGerald's hypothesis had a major advantage over all the others. It was impossible to disprove. It said simply that there is a one-dimensional contraction (in the direction of motion) that increases as velocity increases. The catch is that *everything* contracts. If we want to measure the length of an object that is moving very fast compared to the speed of light, we have to catch up with it first, and when we do, according to the theory, the measuring stick that we are carrying with us also contracts. If the object measured seventeen inches at rest, it still would measure seventeen inches. Nor would anything look contracted because the lenses in our eyes also would contract, distorting them just enough to make everything look normal.

One year later a Dutch physicist, Hendrik Antoon Lorentz, while working on another problem, independently arrived at FitzGerald's hypothesis. Lorentz, however, expressed his discovery in rigorous mathematical terms. This, of course, upgraded FitzGerald's hypothesis to a position of respectability and it began to gain a surprising degree of acceptance, considering its fantasy-like quality. Lorentz's mathematical formulations of the FitzGerald-Lorentz contraction became known as the Lorentz transformations.

The stage was now set. All of the scenery was in place. The failure to detect the ether. The Michelson-Morley experiment.* The constancy of the speed of light. The FitzGerald-Lorentz contractions. The Lorentz transformations. These are the facts that continued to confuse physicists at the beginning of the century. All of them but Albert Einstein. When he looked at these pieces of scenery, what his beginner's mind saw was the special theory of relativity.

*It is said that the reasoning process by which Einstein discovered the special theory of relativity did not include the results of the Michelson-Morley experiment. However, the results of this well-publicized experiment were "in the air" for eighteen years prior to Einstein's paper on special relativity (1905) and they led to the Lorentz transformations which became central to the mathematical formalism of special relativity.

1

✺

Special Nonsense

*Einstein's first professional act, upon reviewing the facts, was the equiva-*lent of saying, "But the Emperor's not wearing any clothes!" except what he said was, "The ether does not exist."[1] The first message of the special theory of relativity is that since the ether is undetectable and, in effect, useless, there is no reason to continue to search for it. It is undetectable because every attempt to measure it or determine its quality, culminating with the Michelson-Morley experiment, failed utterly even to indicate its presence. It is useless because light propagation can be envisioned as the propagation of energy through *empty space (in vacuo)* according to Maxwell's field equations as well as it can be envisioned as a disturbance of the ether medium. Einstein stated clearly what already was implicit in Maxwell's equations. (Maxwell was the discoverer of the electromagnetic field.) "The electromagnetic fields," he wrote, "are not states of a medium [the ether] and are not bound down to any bearer, but they are independent realities which are not reducible to anything else . . ."[2] This assertion was supported by the inability of physicists to detect the ether.

With this statement, Einstein brought to a close the illustrious history of mechanics, the idea that physical events are explicable in terms of things. Classical mechanics is the story of objects and forces between them. It was a remarkable break from a three-century-old tradition to assert blatantly, in the early 1900s, that electromagnetic fields involve no object whatever, that they are not states of the ether medium, but "ultimate, irreducible realities"[3] in themselves. Henceforth, as in quantum mechanics, there would be no concrete imagery associated with physical theory.

Both relativity and quantum theory heralded the unprecedented remoteness from experience which has characterized physical theory ever since. In fact, the trend is continuing. As though governed by an inexorable law, physics is becoming more and more abstract as it covers wider and wider tracts of experience. Only the future will tell if this trend is reversible.

The second victim of Einstein's inability to see clothes that weren't there was absolute nonmotion. Why should we make one particular frame of reference "privileged"[4] in respect to all others by saying that it alone absolutely is not moving? It may be desirable theoretically, but since such a frame of reference does not constitute a part of our experience, it should be disregarded. It is "intolerable"[5] to place in a theoretical structure a characteristic which has no corresponding characteristic in our system of experience.

In one stroke, Einstein eliminated the two major physical and philosophical blocks to a radically new way of perceiving reality. With no ether and no concept of absolute motion to confuse the situation, the situation became much simpler.

Einstein's next step was to confront the puzzle which had come to light (no pun) in the Michelson-Morley experiment, namely, the constancy of the speed of light. How could the speed of light *always* be 186,000 miles per second regardless of the state of motion of the observer?

In an ingenious mental turnaround, Einstein turned this puzzle into a postulate! Instead of worrying, for the moment, about how it can happen, he simply accepted the experimentally irrefutable fact that it *does* happen. This evident (to us) recognition of the obvious was the first step in a logical process, which, once set in motion, was to explain not only the puzzle of the constant speed of light, but a great deal more.

The puzzle of the constancy of the velocity of light became the principle of the constancy of the velocity of light. The principle of the constancy of the velocity of light is the first foundation stone of the special theory of relativity.

The principle of the constancy of the velocity of light is that whenever we make a measurement of the velocity of light, regardless of whether we are in motion or at rest relative to the light source, we always get the same result. The speed of light is invariably 186,000 miles per second.* This is what Michelson and Morley discovered in their famous experiment.

From the point of view of classical mechanics, the principle of the constancy of the velocity of light makes no sense at all. In fact, it conflicts violently with common sense. Before Einstein, the totalitarian grasp of "common sense" held the constancy of the speed of light to the status of a paradox. (Whenever we bump into the limits of our self-imposed cognitive reality, the result is always paradox.) It took a pure beginner's mind, such as Albert Einstein's, to accept that if what is, is (the constancy of the velocity of light), then *common sense must be wrong*.

The most important victim of Einstein's beginner's mind was the whole structure of classical (Galilean) transformations, that sweet but illusory fruit of a common sense anchored in macroscopic dimensions and velocities. To give up common sense is not an easy task. Einstein was the first person to do it in such a wholesale manner that his perception of the very nature of space and time changed radically. Moreover, when all was said and done, Einstein's vision of space and time turned out to be more useful than that of common sense.

The second foundation stone of the special theory of relativity is the principle of relativity. When Einstein dismissed the idea of absolute

---

*In a vacuum. The speed of light changes in matter depending upon the index of refraction of the matter.

nonmotion, his theory became, *ipso facto,* a theory of relativity. Since there was no better principle of relativity to be had than Galileo's, Einstein simply borrowed it, but first, of course, he brought it up to date.

Galileo's principle of relativity says that the laws of mechanics (such as the laws governing falling bodies) that are valid in one frame of reference are valid in all frames of reference that move uniformly (without jerkiness) in relation to it. Another way of saying the same thing is that it is impossible to determine, by doing experiments involving the laws of mechanics, whether or not our frame of reference is moving or at rest in relation to another frame of reference in which the laws of mechanics also are valid.

Einstein expanded the Galilean relativity principle to include *all* the laws of physics, and not just the laws of classical mechanics. In particular he included the laws governing electromagnetic radiation, which were unknown in Galileo's time.

Einstein's updated principle of relativity, then, is that all the laws of nature are exactly identical in all frames of references that move uniformly relative to each other and that, therefore, there is no way of distinguishing absolute uniform motion (or nonmotion).

In short, the two foundation stones of the special theory of relativity are the principle of the constancy of the velocity of light (the Michelson-Morley experiment) and the principle of relativity (Galileo). Said more specifically, the special theory of relativity rests upon these two postulates:

(1) The velocity of light in a vacuum is the same in all frames of reference (for all observers) moving uniformly, relative to each other, and

(2) All laws of nature are the same in all frames of reference moving uniformly, relative to each other.

Of these two postulates, the first one, the principle of the constancy of the velocity of light, is the troublemaker. There is no way that it and the classical transformation laws both can be true. According to the classical transformation laws (and common sense) the speed of light must be its velocity as it is emitted from a source plus or minus the velocity of the observer, if the observer is moving toward the source or away from the source. According to experiment, the speed of light remains constant regardless of the state of motion of the observer. Common sense and experimental findings are in violent disagreement.

Einstein's beginner's mind told him that, since we cannot argue with what is (the experimental evidence), then our common sense must be wrong. With this decision to disregard common sense and to base his new theory on the only clothes he could see that the emperor *was* wearing (the constant speed of light and the principle of relativity) Einstein stepped boldly into the unknown, in fact, into the unimaginable. Already on new territory, he proceeded to explore where no person had gone before.

How could it be that to every observer the speed of light is the same regardless of their state of motion? To measure speed, it is necessary to use a clock and ruler (a rigid rod). If the speed of light as measured by an observer at rest relative to a light source is the same as the speed of light as measured by an observer in motion relative to the source, then it must be that, somehow, *the measuring instruments* change from one frame of reference to the other in just such a way that the speed of light always appears to be the same.

The speed of light appears constant because the rods and clocks used to measure it vary from one frame of reference to another depending upon their *motion*. In short, to an observer at rest, a moving rod changes its length and a moving clock changes its rhythm. At the same time, to an observer traveling along with a moving rod and clock, there is no apparent change at all in length or rhythm. Therefore, both observers measure the speed of light to be the same, and neither can detect anything unusual in the measurement or in the measuring apparatus.

This is very similar to the case of the Michelson-Morley experiment. According to FitzGerald and Lorentz, the arm of the interferometer that faces into the ether wind (now dismissed from our theory) is shortened by the pressure of the ether wind. Therefore, the light that travels the interferometer arm facing into the "ether wind" has less distance to travel and more time to do it in than does the light traveling the other arm. As a result, the speed of light traveling both arms appears to be the same. This is what the Lorentz transformations describe. Come to think of it, *the Lorentz transformations can be used to describe contractions due to motion* as well as contractions due to a fictitious ether wind.

FitzGerald and Lorentz imagined that rigid rods were compressed under the pressure of the ether wind, but according to Einstein, it is *motion itself* that causes contraction, and, in addition, time dilation.

Here is another way of looking at it. A "constant velocity of light" is exactly what would result if moving measuring rods became shorter and moving clocks ran more slowly because a moving observer would measure the speed of light with a shorter measuring rod (less distance for the light to travel) and a slower clock (more time to do it in) than an observer at rest. Each observer, however, would consider his own rod and clock to be quite normal and unimpaired. Therefore, both observers would find the speed of light to be 186,000 miles per second and both of them would be puzzled by this fact if they were still bound by the classical transformation laws.

These were the initial fruits of Einstein's basic assumptions (the principle of the constancy of the velocity of light and the principle of relativity): First, a moving object appears to contract in its direction of motion and become shorter as its velocity increases until, at the speed of light, it disappears altogether. Second, a moving clock runs more slowly than a clock at rest and continues to slow its rhythm as its velocity increases until, at the speed of light, it stops running altogether.

These effects only appear to a "stationary" observer; one who is at rest relative to the moving clock and rod. They do not appear to an observer who is traveling along with the clock and rod. To make this clear, Einstein introduced the labels "proper" and "relative." What we see when we observe our stationary rod and our stationary clock, if we ourselves are stationary, is their *proper* length and *proper* time.

("Proper" means "one's own.") Proper lengths and proper times always appear normal. What we see if we are stationary and observe a rod and a clock traveling very fast relative to us is the *relative* length of the moving rod and the *relative* time of the moving clock. The relative length is always shorter than the proper length, and the relative time is always slower than the proper time.

The time that you see on your own watch is your proper time, and the time that you see on the watch of the person moving past you is the relative time (which appears to you—not to the person moving past you—to run more slowly). The length of the measuring rod in your own hand is its proper length, and the length of the measuring rod in the hand of the person moving past you is its relative length (which appears to you—but not to the other person—to be shorter). From the point of view of the person moving past you, he is at rest, you are moving, and the situation is reversed.

Suppose that we are aboard a spacecraft outward bound on an exploration. We have made arrangements to press a button every fifteen minutes to send a signal back to earth. As our speed steadily increases our earthbound colleagues notice that instead of every fifteen minutes, our signals begin to arrive seventeen minutes apart, and then twenty-five minutes apart. After several days, our colleagues, to their distress, find that our signals arrive every two days. As our velocity continues to increase our signals become years apart. Eventually, generations of earthlings come and go between our signals.

Meanwhile, on the spacecraft, we are entirely unaware of the predicament back on earth. As far as we are concerned, everything is proceeding according to plan, although we are becoming bored with the routine of pressing a button every fifteen minutes. When we return to earth, a few years older (our proper time) we may find that we have been gone, according to earth time, for centuries (their relative time). Exactly how long depends upon how fast we have been going.

This scene is not science fiction. It is based upon a well-known (to physicists) phenomenon called the Twin Paradox of the special theory of relativity. Part of the paradox is that one twin remains on earth while the other goes on a space voyage and returns younger than his brother.

There are many examples of proper time and relative time. Suppose that we are in a space station observing an astronaut who is traveling at a speed of 161,000 miles per second relative to us. As we watch him, we notice a certain sluggishness in his movements, as though he were moving in slow motion. We also notice that everything in his spaceship also seems to function in slow motion. His rolled cigarette, for example, lasts twice as long as one of ours.

Of course, part of his sluggishness is due to the fact that he is fast increasing the distance between us, and with each passing moment, it takes the light from his spaceship longer to reach us. Nonetheless, after making allowances for the travel time of the light involved, we find that the astronaut still is moving more slowly than usual.

However, to the astronaut, it is we who are zipping past him at 161,000 miles per second, and after he makes all the necessary allowances, he finds that it is we who are sluggish. Our cigarette lasts twice as long as his.

This situation could be the ultimate illustration of how the grass is always greener on the other side. Each man's cigarette lasts twice as long as the other's. (Unfortunately, so does each man's trip to the dentist.)

The time that we ourselves experience and measure is our proper time. Our cigarette lasts the normal length of time. The time that we measure for the astronaut is the relative time. His cigarette appears to last twice as long as ours because his time passes twice as slowly. The situation is similar regarding proper lengths and relative lengths. From our point of view, the astronaut's cigarette, provided that it is pointing in the direction that his spaceship is moving, is shorter than our own cigarette.

The other side of the coin is that the astronaut sees himself as stationary and his cigarettes as normal. He also sees us as traveling at 161,000 miles per second relative to him, and our cigarettes as shorter than his and slower burning.

Einstein's theory has been substantiated in many ways. All of them verify it with awesome accuracy.

The most common verifications of time dilation come from high-energy particle physics. A very light elementary particle, called a muon (pronounced moo'on), is created at the top of the earth's atmosphere by the collision of protons (one form of "cosmic radiation") and air molecules. We know from experiments in which muons are created in accelerators that they live a very short time. By no means do they live long enough to reach the earth from the upper atmosphere. Long before the time it takes to traverse this distance, they should decay spontaneously into other types of particles. Yet this does not happen because we detect them in abundance here at the earth's surface.

Why do the muons created by cosmic radiation live longer, in fact, *seven times longer* than those muons created in the laboratory? The answer is that the muons produced by collisions of cosmic radiation and air molecules travel much faster than any muons that we can create experimentally. Their velocity is approximately 99 percent of the speed of light. At that speed, time dilation is quite noticeable. They do not live longer than usual from *their* point of view, but from our point of view they live seven times longer than they would at slower velocities.

This is true not only of muons, but of almost all subatomic particles, and there are many of them. For example, pions (pie'ons), another type of subatomic particle, which move at 80 percent of the speed of light, live, on the average, 1.67 times as long as slow pions. The special theory of relativity tells us that the intrinsic lifetime of these high-speed particles does not increase, but that their relative rate of time flow slows down. The special theory of relativity also made the calculation of these phenomena possible long before we had the technical capability to create them.

In 1972, four of the most accurate atomic clocks available were put aboard an aircraft and flown around the world. At the end of the trip, they were found to be slightly behind their stationary, earth-bound counterparts with which they were synchronized before the flight.* The next time that you fly, remember that, even if minutely, your watch is running slower, your body had more mass, and, if you stand facing the cockpit, you are thinner.

According to the special theory of relativity, a moving object appears to contract in the direction of motion as its velocity increases. James Terrell, a physicist, has demonstrated mathematically that this phenomenon is something like a visual illusion, and, in fact, is analogous to a projection of the real world onto the wall of Plato's cave.[6]

Plato's famous parable of the cave describes a group of people who are chained inside a cave in such a way that they can see only the shadows on the wall of the cave. These shadows are the only world that these people know. One day one of these people escapes into the world outside the cave. At first he is blinded by the sunlight, but when he recovers, he realizes that *this* is the real world, and what he previously considered to be the real world was, in fact, only the projection of the real world onto the wall of the cave. (Unfortunately, when he returned to the people who still were chained inside the cave, they thought he was mad.)

Figure A, on the next page, depicts a view looking down on the top of our head and the top of a sphere. The lines connect our eyes with points on either side of the sphere. If we are far enough away from the sphere, the distance between these points is almost equal to the diameter of the sphere. Figure A is drawn as if the artist were looking down on the top of our head, our eyes, and the sphere.

The first step in Terrell's explanation is to draw lines downward (back into the page) from each of the two points on the sphere to a screen directly below the sphere. Figure B is a side view showing the two points, the lines that we have drawn downward, and the screen. (If you hold this book directly in front of you, your eyes are in the same position relative to the dotted lines as the eyes drawn in Figure A.)

To understand Terrell's explanation, suppose that the sphere is

*The clocks were flown around the world each way (east and west). Both general relativistic and special relativistic effects were noted. (J. C. Hafele and R. E. Keating, *Science,* vol. 177, 1972, pp. 168ff.)

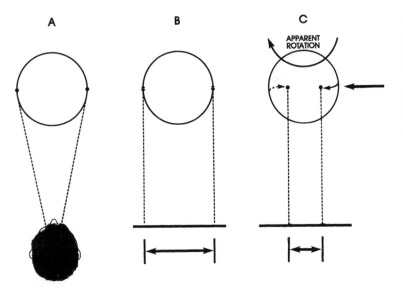

moving *very fast* relative to the speed of light from right to left. If the sphere moves fast enough, some very interesting things happen. For example, before the light from the point on the far left edge of the sphere can reach us, the ball moves in front of it, blocking it from our vision! The reverse happens on the far right. The ball moves out from between us and the light signals originating from points that used to be on the "back" side of the ball. These signals now are visible to us, while the signals coming from the point that used to be on the leading edge of the ball now are blocked by the ball itself as it moves to the left. The effect of this is an illusion of sorts. What we see is the same thing that we would see if someone had *rotated* the ball around its axis!

Look what happened to the distance between the two points as projected on the screen. It is considerably less than when we started. The equations in the special theory of relativity (the Lorentz transformations) which show a contraction due to motion describe these *projections*. (Is this beginning to sound like Plato's cave?)

The fact that the ball, by moving fast enough, gets in the way of some of its own light signals and out of the way of others causes the ball to appear to rotate. This causes the *projected* distance between any two points on it which are aligned with the direction of motion to decrease, just as if someone really had rotated the ball. The faster the ball moves, the more it appears to "rotate," and the closer together come the points projected on the screen. It is the projection that contracts. Instead of "screen" substitute "view of the ball from our frame of reference" and we have the Terrell explanation of relativistic contraction.

As yet, no analogous explanations have been found for the time dilation that accompanies moving clocks or the increase of mass that accompanies moving objects, but the effort, relatively speaking, is young.

The special theory of relativity shows that the mass of a moving object increases as the velocity of the object increases. Newton would not have hesitated to call this nonsense, but then Newton's experience was limited to velocities that are quite slow compared to the speed of light.

Classic physics tells us that a specific amount of force is required to increase the velocity of a moving object by a given amount, for example, one foot per second. Once we know what that amount of force is, whenever we want to increase the velocity of that particular object by one foot per second, all we need do is apply that amount of force to it. If the object has a velocity of 100 feet per second, that specific amount of force will increase its velocity to 101 feet per second. According to Newton's physics, the same amount of force that increases the velocity of an object from 100 feet per second to 101 feet per second also will increase the velocity of the same object from 8,000 feet per second to 8,001 feet per second.

The problem is that Newtonian physics is wrong. It takes much more force to increase by one foot per second the velocity of an object moving at 8,000 feet per second than it takes to increase by one foot per second the velocity of the same object moving at 100 feet per second.

That is because a faster-moving object has more kinetic energy (energy of motion). This additional energy makes it behave exactly as if it had more mass. A given amount of force applied for a given amount of time will accelerate a force applied for the same amount of time to an entire train. Of course, this is because an entire train has more mass than a single car.

When particles travel at velocities that are fast relative to the speed of light, their high kinetic energy makes them behave as though they have more mass than they have at lower velocities. In fact, the special theory of relativity shows that the effective mass of a moving object *does* increase with velocity.

Since most subatomic particles travel at different velocities, each one of them can have many different relative masses. Therefore, physicists have calculated the "rest mass" of each particle. The rest mass of a subatomic particle is its mass when it is not moving. Subatomic particles are never really at rest, but these calculations provide a uniform method of comparing their masses. This is necessary since, as the velocity of a particle approaches the speed of light, its relative mass depends upon how fast it is moving.

Einstein's discovery that moving clocks change their rhythm led to some spectacular revisions in the way that we see the world. It showed that there is no "universal" time that permeates the universe. There are only proper times associated with various observers. The proper time of each observer is different, unless two of them happen to be at rest relative to each other. If the universe has a heart beat, its rate depends upon the hearer.

The special theory of relativity shows that two events which happen at the same time in one frame of reference may occur at different times when seen from another frame of reference. To illustrate this point, Einstein used one of his famous thought experiments.

A thought experiment is a mental exercise. It has the advantage of requiring no apparatus other than the mind, which frees it from the practical limitations of laboratory experiments. Most physicists accept the use of thought experiments as a valid theoretical tool, provided they are satisfied that if the experiment could be performed, the results of the actual experiment would be the same as those of the thought experiment.

Suppose that we are in a moving room. The room is moving with a uniform velocity. Exactly in the center of the room is a light bulb which flashes periodically. The room is made of glass so that an outside observer can see what happens inside.

At the precise moment that we pass an outside observer, the light flashes. The question is, is there any difference between what we see inside the moving room and what the outsider observer sees? According to the special theory of relativity, the answer is an extraordinary, concept-shattering *Yes*. There is a big difference.

Inside the room, we see the bulb flash and we see the light spread out in all directions at the same speed. Since the walls of the room are equidistant from the bulb, we see the light strike the forward wall and rearward wall of the room simultaneously.

The outside observer also sees the flash, and he also sees the light propagate in all directions at the same speed. However, in addition, he sees that the room is moving. From his point of view the forward wall tries to escape the approaching light while the rearward wall rushes to meet it. Therefore, to the outside observer, the light reaches the rearward wall before it reaches the forward wall. If the speed of the room is small compared to the speed of light, the light reaches the rearward wall only slightly ahead of the forward wall. Nonetheless, the light reaches the rearward and forward walls in a one-two order, and not at the same time.

Although both of us observed the same two events, the light striking the forward wall and the light striking the rearward wall, we each have different stories to tell. To us, inside the room, the two events were simultaneous. To the outside observer, one event came first and the other event came later.

Einstein's revolutionary insight was that events which are simultaneous for one observer may occur at different times for another observer depending upon their relative motion. Put another way, two events, one of which occurs before the other as seen from the frame of reference of one observer, may occur at the same time when seen from the frame of reference of another observer. One observer uses the words "sooner" and "later." The other observer uses the word "simultaneous," even though both of them are describing the same two events.

In other words, "sooner," "later," and "simultaneous" are local terms. They have no meaning in the universe at large unless they are tied down to a specific frame of reference. What is "sooner" in one frame of reference may be "later" in another frame of reference and "simultaneous" in a third.*

The mathematics which translate what an observer in one frame of reference sees into what an observer in another frame of reference sees are the Lorentz transformations. Einstein adopted the Lorentz transformations—which are a set of equations—virtually intact.

No one before Einstein got these startling results from this simple type of thought experiment because no one before Einstein had the audacity to postulate something as outrageous as the principle of the constancy of the velocity of light. No one had the audacity to postulate something as outrageous as the principle of the constancy of the velocity of light because the principle of the constancy of the velocity of light completely and unequivocally contradicts common sense; specifically, common sense as represented by the classical transformation laws. The classical transformation laws are so embedded in our everyday experience that it simply never occurred to anybody to question them.

Even when the Michelson-Morley experiment produced results that were incompatible with the classical transformation laws, no beginner's mind but Einstein's conceived that the classical transformation laws might be wrong. Only Einstein suspected that at very high velocities, velocities far faster than those that we encounter through

---

*This is only true for events that are space-like separated. For time-like separated events the relation earlier-later is preserved for all observers. Time-like separated events can never appear simultaneous in any frame of reference moving with a velocity less than c. (Space-like separation is explained later.)

our senses, the classical transformation laws do not apply. This is not to say that they are incorrect. At low velocities (compared to 186,000 miles per second) contraction and time dilation are not detectable sensorily. In this limited situation, the classical transformations are a good guide for practical experience. After all, we do reach the top of an escalator faster if we walk.

If we do the moving-room experiment with sound instead of light, we do not get the special theory of relativity. We get a confirmation of the classical transformation laws. There is no principle of the constancy of the velocity of sound because the velocity of sound is not constant. It varies depending upon the motion of the observer (hearer) as dictated by common sense. The important word here is "dictated."

We live out our lives in a limited situation of low velocities where the speed of sound (about 700 miles per hour) seems "fast." Therefore, our common sense is based upon our experiences in this limited environment. If we want to expand our understanding beyond the limitations of this environment, it is necessary to drastically rearrange our conceptual constructs. This is what Einstein did. He was the first person to see that this is what had to be done in order to make sense of such impossible experimental findings as the constancy of the velocity of light for each and every person who measures it, regardless of their states of motion.

This led him to turn the puzzle of the constancy of the velocity of light into the principle of the constancy of the velocity of light. In turn, that led him to the conclusion that, if the velocity of light really is constant for all observers, then the measuring instruments used by different observers in different states of motion somehow must vary so that all of them give the same result. By a stroke of luck, Einstein discovered that these same variances were expressed in the equations of the Dutch physicist Hendrik Lorentz, and so he borrowed them. Lastly, the fact that moving clocks change their rhythm led Einstein to the inescapable conclusion that "now," "sooner," "later," and "simultaneous" are *relative* terms. They all depend upon the state of motion of the observer.

This conclusion is precisely the opposite of the assumption upon which Newtonian physics is based. Newton assumed, as did we all, that there is one clock ticking off the seconds by which the entire universe grows older. For every second of time that passes in this corner of the universe, one second of time passes also in every other corner of the universe.

According to Einstein, this is incorrect. How can anyone say when it is "now" throughout the universe? If we try to designate "now" by the occurrence of two simultaneous events (like my arrival at the doctor's office and my watch indicating 3 o'clock), we find that an observer in another frame of reference sees one of our events happening before the other. Absolute time, wrote Newton, "flows equally . . . ,"[7] but he was wrong. There is no single time which flows equally for all observers. There is no absolute time.

The existence of one ultimate flow of time throughout the physical universe, which we all tacitly acknowledged, turned out to be another piece of clothing that the Emperor wasn't wearing.

Newton made one more mistake in this regard. He said that time and space were separate. According to Einstein, time and space are not separate. Something cannot exist at some place without existing at some time, and neither can it exist at some time without existing at some place.

Most of us think of space and time as separate because that is the way that we think that we experience them. For example, we seem to have some control over our position in space, but none at all over our position in time. There is nothing that we can do about the flow of time. We can choose to stand perfectly still, in which case our position in space does not change, but there is no way that we can stand still in time.

This notwithstanding, there is something very elusive about "space" and especially about "time"; something that prevents us from "resting our accounts with them prematurely." Subjectively, time has a fluid quality which much resembles a running brook; sometimes bubbling past in a furious rush, sometimes slipping by quietly unnoticed, and sometimes lying languid, almost stationary, in deep pools. Space, too, has an ubiquitous quality about it which belies the common notion that it serves only to separate things.

William Blake's famous poem reaches out toward these intangible qualities:

> To see a World in a Grain of Sand
> And a Heaven in a Wild Flower,
> Hold Infinity in the palm of your hand
> And Eternity in an hour.

(Its title, by no coincidence is "Auguries of Innocence.")

The special theory of relativity is a physical theory. Its concern is with the mathematically calculable nature of reality. It is *not* a theory of subjectivity. Although it shows that the appearances of physical reality may vary from one frame of reference to another, it is a theory about the unchanging (physicists say "invariant") aspect of physical reality. Nonetheless, the special theory of relativity was the first mathematically rigorous physical theory to explore areas whose expression previously had been the domain of poets. Like any concise and poignant *re-presentation* of reality, the theories of relativity *are* poetry to mathematicians and physicists. However, Albert Einstein's enormous public renown perhaps was due in part to a shared intuition that he had something profoundly relevant to say about space and time.

What Einstein had to say about space and time is that there is no such thing as space *and* time; there is only space-time. Space-time is a continuum. A continuum is something whose parts are so close together, so "arbitrarily small," that the continuum really cannot be broken down into them. There are no breaks in a continuum. It is called a continuum because it flows continuously.

For example, a one-dimensional continuum is a line drawn on a wall. Theoretically we might say that the line is comprised of a series of points, but the points are each infinitely close to one another. The result is that the line flows continuously from one end of it to the other.

An example of a two-dimensional continuum is the wall. It has two dimensions, length and width. Similarly, all of the points on the wall are in contact with other points on the wall, and the wall itself is a continuous surface.

A three-dimensional continuum is what we commonly call "space." A pilot flying his airplane navigates in a three-dimensional continuum. To give his location he must state, for example, not only how far north and how far east of a given point he is, but he also must report his altitude. The airplane itself, like all things physical, is three-dimensional. It has a width, a height, and a depth. This is why mathematicians call our reality (their reality, too) three-dimensional.

According to Newtonian physics, our three-dimensional reality is separate from, and moves forward in, a one-dimensional time. Not so, says the special theory of relativity. Our reality is *four-dimensional,* and the fourth dimension is time. We live, breathe, and exist in a four-dimensional space-time continuum.

The Newtonian view of space and time is a *dynamic* picture. Events *develop* with the passage of time. Time is one-dimensional and *moves* (forward). The past, present, and future happen in that order. The special theory of relativity, however, says that it is preferable, and more useful, to think in terms of a *static,* nonmoving picture of space and time. This is the space-time continuum. In this static picture, the space-time continuum, events do not develop, they just are. If we could view our reality in a four-dimensional way, we would see that everything that now seems to unfold before us with the passing of time, already exists *in toto,* painted, as it were, on the fabric of space-time. We would see all, the past, the present, and the future with one glance. Of course, this is only a mathematical proposition (isn't it?).

Don't worry about visualizing a four-dimensional world. Physicists can't do it, either. For the moment, just assume that Einstein might be right since the evidence so far suggests that he is. His message is that space and time are related in an intimate manner. For lack of a better way of saying it, he expressed this relationship by calling time a fourth dimension.

"Fourth dimension" is a translation from one language to another. The original language is mathematics and the second language is English. The problem is that there is simply no way of precisely expressing what the first language says in terms of the second language. Therefore, "time as a fourth dimension" is merely a label that we give to a *relationship*. The relationship in question is the relationship between space and time as it is expressed mathematically in the theories of relativity.

The relationship between space and time that Einstein discovered is similar to the relationship between the sides of a right triangle which Pythagoras the Greek (a contemporary of Confucius) discovered about 550 B.C.

A right triangle is a triangle that contains a right angle. A right angle is formed whenever two perpendicular lines intersect. Below is a right triangle. The side of a right triangle that is opposite the right angle is called the hypotenuse ($\overline{\text{hi}}$ pot' n $\overline{\text{oos}}$'). The hypotenuse is always the longest side of a right triangle.

Pythagoras discovered that as long as we know the length of the two shorter sides of a right triangle, we can calculate the length of the longest side. This relationship, expressed mathematically, is the Pythagorean theorem: The first leg squared plus the second leg squared equals the hypotenuse squared.

A hypotenuse of a given length can be calculated from many different combinations of shorter legs. In other words, there are many combinations of different-size legs that all calculate to have the same hypotenuse.

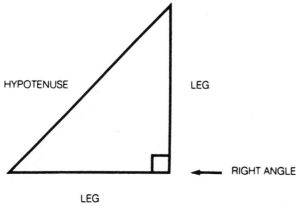

For example, the first leg might be very short and the second leg very long.

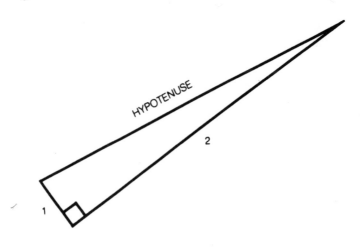

or the other way round,

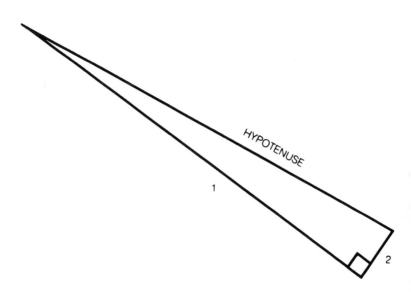

or anything in between.

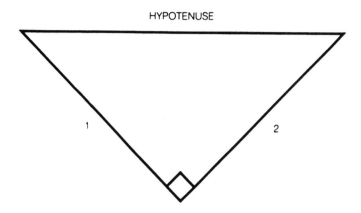

HYPOTENUSE

1   2

If we substitute "space" for one of the legs of a right triangle, "time" for the other leg, and "space-time interval" for the hypotenuse, we have a relationship which is conceptually analogous to the relationship between space, time, and the space-time interval described in the special theory of relativity.* The space-time interval between two events is an absolute. It never varies. It can appear differently to observers in different states of motion, but it is, itself, invariant. The special theory of relativity shows how observers in different frames of reference can observe the same two events and calculate the space-time interval between them. The answer that all of the observers get will be the same.

One observer may be in a state of motion such that for him there

---

*The Pythagorean theorem is $c^2 = a^2 + b^2$. The equation for the space-time interval in the special theory of relativity is $s^2 = t^2 - x^2$. The Pythagorean theorem describes properties in Euclidean space. The equation for the space-time interval describes properties in Minkowski's flat space-time (Euclidean and non-Euclidean space are discussed in the next chapter). There are other differences as well, but the fundamental relationship between space, time, and the space-time interval is very similar to the relationship expressed in the Pythagorean theorem between the three sides of a right triangle.

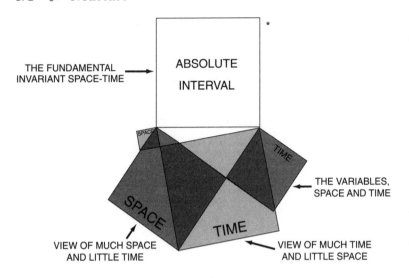

THE FUNDAMENTAL INVARIANT SPACE-TIME → ABSOLUTE INTERVAL

SPACE | TIME

THE VARIABLES, SPACE AND TIME

SPACE | TIME

VIEW OF MUCH SPACE AND LITTLE TIME

VIEW OF MUCH TIME AND LITTLE SPACE

is a time and a distance involved between the two events, and another observer may be in a state of motion such that his measuring devices indicate a different distance and a different time between the events, but the space-time interval between the two events does not vary. For example, the space-time interval, the absolute separation, between two exploding stars is the same whether it is viewed from a slow-moving frame of reference, like a planet, or from a fast-moving frame of reference, like a speeding rocket.

Let us return to our experiment with the moving glass room. Although we inside the room saw the light strike the rearward and forward walls simultaneously, the outside observer saw the light strike the rearward wall before it reached the forward wall. Nevertheless, by using a Pythagorean-like equation, into which we and the outside observer feed our time and distance measurements, we both get the same space-time interval between the events.

Actually, this Pythagorean-like relationship was the discovery of

*Thanks to Guy Murchie who drew the original version of this drawing in his fine book, *Music of the Spheres,* New York, Dover, 1961.

Einstein's mathematics teacher, Hermann Minkowski, who was inspired by his most famous student's special theory of relativity. In 1908 Minkowski announced his vision this way:

> Henceforth space by itself, and time by itself, are doomed to fade away into mere shadows, and only a kind of union of the two will preserve an independent reality.[8]

Minkowski's mathematical explorations of space and time were both revolutionary and fascinating. Out of them came a simple diagram of space-time showing the mathematical relationship of the past, present, and the future. Of the wealth of information contained in this diagram, the most striking is that all of the past and all of the future, for each individual, meet and forever meet, at one single point, *now*. Furthermore, the *now* of each individual is specifically located and will never be found in any other place than *here* (wherever the observer is at).

Sixty-three years before Ram Dass's great book, *Be Here Now,* established the watchwords of the awareness movement, Hermann Minkowski proved that, in physical reality, no choice exists in the matter (pun?). Unfortunately for physicists, the realization is not always the experience. Nonetheless, after two thousand years of use in the East, being here now, the beginning step in meditation, received the validation of western science via Minkowski's rigorous mathematical confirmation of it, inspired by the special theory of relativity.

The last and the most famous aspect of the special theory of relativity is the revelation that mass is a form of energy, and that energy has mass. In Einstein's words, "Energy has mass and mass represents energy."[9]

Although this sounds shocking in one sense, in the sense that we have believed ever so long that matter, stuff, is different from energy just as the body is different from the mind (another form of the same theory), in another sense, it sounds surprisingly natural. The matter-energy dichotomy goes back at least as far as the Old Testament. Genesis portrays man as a sort of ceramic creation. God scoops up a handful of clay (matter) and breathes life (energy) into it. The Old Testament is a product of the western world (or the other way round). Physics also is a product of the western world.

In the East, however, there never has been much philosophical or religious (only in the West are these two separate) confusion about matter and energy. The world of matter is a relative world, and an illusory one: illusory not in the sense that it does not exist, but illusory in the sense that we do not see it as it really is. The way it really is cannot be communicated verbally, but in the attempt to talk around it, eastern literature speaks repeatedly of dancing energy and transient, impermanent forms. This is strikingly similar to the picture of physical reality emerging from high-energy particle physics. Buddhist literature does not speak of learning new things about reality, but about removing veils of ignorance that stand between us and what we already *are*. Perhaps this accounts for the fact that the preposterous claim that mass is only a form of energy is unexpectedly palatable.

The formula which expresses the relationship of mass to energy is the most famous formula in the world: $E = mc^2$. The energy contained in a piece of matter is equal to the mass of the matter multiplied by an extraordinarily large number, the speed of light squared. This means that even the tiniest, the very tiniest particle of matter has within it a tremendous amount of concentrated energy.

Although Einstein didn't know it at the time, he discovered the secret of stellar energy. Stars continuously convert matter into energy. It is because of the very large ratio of energy released to matter consumed that stars can continue to burn through countless millennia.

At the center of a star, hydrogen atoms, the primordial "stuff" of the physical world, are squeezed together so tightly by the enormous gravitational force of the star's dense mass that they fuse together, making a new element, helium. Every four hydrogen atoms become one helium atom. However, the mass of one helium atom is not the same as the mass of four hydrogen atoms. It is slightly less. This small difference in mass is released as radiant energy—heat and light. The process of fusing lighter elements into heavier elements is called, of course, fusion. The fusion of hydrogen into helium causes a hydrogen explosion. In other words, a (young) burning star literally is one huge, continuously exploding hydrogen bomb.*

The formula $E = mc^2$ also resulted in the atomic bomb. Atomic bombs and atomic reactors obtain energy from mass by the process of fission, which is the opposite of fusion. Instead of fusing smaller atoms into larger ones, the process of fission splits atoms of uranium, which are quite large, into atoms which are smaller.

This is done by firing a subatomic particle, a neutron, at an atom of uranium. When the neutron hits the uranium atom, it splits it into lighter atoms, but the mass of these smaller atoms together is less than the mass of the parent atom of uranium. The difference in mass explodes into energy. This process also produces additional neutrons which fly off to strike other uranium atoms, creating more fissions, more light atoms, more energy, and more neutrons. The whole phenomenon is called a chain reaction. An atomic bomb is an uncontrolled chain reaction.

A hydrogen (fusion) bomb is produced by detonating an atomic (fission) bomb in the midst of hydrogen. The heat from the atomic explosion (in place of the heat of friction caused by gravity) fuses hydrogen atoms into helium atoms and releases heat in the process which fuses together more hydrogen atoms, releasing more heat, and so on. There is no limit to the size of a potential hydrogen bomb, and it is constructed from the most plentiful element in the universe.

For better or worse, a major revelation of the special theory of relativity is that mass and energy are different forms of the same thing. Like space and time, they are not separate entities. There is no qualitative difference between mass and energy; there is only mass-energy. Mathematically, this discovery meant that the two conservation laws of mass and energy could be replaced by a single conservation law of mass-energy.

---

*As its hydrogen becomes exhausted, a star begins to fuse the helium at its core. Helium fusion is hotter than hydrogen fusion and produces heavier elements, such as neon, oxygen, and carbon, which, in turn, becomes the solar fuel as its helium becomes exhausted.

A conservation law is a simple statement that a quantity of something, whatever it may be, never changes no matter what happens. For example, suppose that there were a conservation law governing the number of guests at a party. If such a thing were true, we would notice that every time a new guest arrived at the party, some other guest would leave. Similarly, every time a guest at the party left, another one would arrive. The rate of guest turnover at the party might be great or small, and the guests might arrive and depart singly or in groups, but in all circumstances the number of guests at the party would remain the same.

The conservation law concerning energy says that the total amount of energy in the universe always has been and always will be the same. We can convert energy from one form to another (like mechanical energy to thermal energy via friction), but the total amount of energy in the universe does not change. Similarly, the law of the conservation of matter says that the total amount of matter in the universe always has been and always will be the same. We can convert matter from one form to another (like ice to water or water to steam), but the total amount of matter in the universe does not change.

When the special theory of relativity combined mass and energy into mass-energy, it also combined the law of the conservation of mass and the law of the conservation of energy into the law of the conservation of mass-energy. The law of the conservation of mass-energy says that the total amount of mass-energy in the universe always has been and always will be the same. Mass may be converted into energy and energy may be converted into mass, but the total amount of mass-energy in the universe does not change.

The sun, the stars, even wood burning in the fireplace, are examples of mass being converted into energy. Physicists who study subatomic particles are so familiar with the concept of exchanging mass for energy and energy for mass that they routinely designate the mass size of particles in terms of their energy content.

In all, there are roughly twelve conservation laws. These simple laws are becoming more and more important, especially in high-energy particle physics, because they are derived from what physicists now believe to be the ultimate principles (latest dance) governing the physical world. These are the laws of symmetry.

The laws of symmetry are pretty much what they sound like. Something is symmetrical if certain aspects of it remain the same under varying conditions. For example, one half of a circle mirrors the other half, no matter how we cut it. Regardless of how we turn a circle, the right half always mirrors the left half. The position of the circle changes, but its symmetry remains.

The Chinese have a similar concept (perhaps the same?). One side of a circle is called "yin" and other side is called "yang." Where there is yin, there is yang. Where there is high, there also is low. Where there is day, there also is night. Where there is death, there also is birth. The concept of yin-yang, which is really a very old law of symmetry, is yet another way of saying that the physical universe is a whole which seeks balance within itself.

The irony of the special theory of relativity, as apparent by now, is that it is not about those aspects of reality that are relative, but about those aspects that are not relative. Like quantum mechanics, its impact on the assumptions of Newtonian physics was shattering. Not because it proved them wrong, but because it proved them to be quite limited. The special theory of relativity and quantum mechanics have propelled us into unimaginably expansive areas of reality, areas about which we literally had not one previous idea.

The assumptions of Newtonian physics correspond to the clothes we always thought that the Emperor was wearing: a universal time whose uniform passage equally affects every part of the universe; a separate space, independent though empty; and the belief that there exists somewhere in the universe a place which stands absolutely still, quiet and unmoving.

Every one of these assumptions has been proven untrue (not useful) by the special theory of relativity. The Emperor wasn't wearing them at all. The only motion in the physical universe is motion relative to something else. There is no separate space and time. Mass and energy are different names for the same thing.

In place of these assumptions, the special theory of relativity provides a new and unified physics. Measurements of distance and duration may vary from one frame of reference to another, but the space-time interval between events never changes.

For all this, however, the special theory of relativity has one shortcoming. It is based on a rather uncommon situation. The special theory of relativity applies only to frames of reference that move uniformly, relative to each other. Most movement, unfortunately, is neither constant nor ideally smooth. In other words, the special theory of relativity is built upon an idealization. It is limited to and premised upon the special situation of uniform motion. That is why Einstein called it the "special," or restricted, theory.

Einstein's vision was to construct a physics that is valid for *all* frames of reference, such as those moving with non-uniform motion (acceleration and deceleration) relative to each other, as well as those moving uniformly relative to each other. His idea was to create a physics which could describe events in terms of *any* frame of reference, no matter how it moves relative to any other frame of reference.

In 1915, Einstein succeeded in achieving the complete generalization of his special theory. He called this achievement the general theory of relativity.

1

✺

General Nonsense

The general theory of relativity shows us that our minds follow different rules than the real world does. A rational mind, based on the impressions that it receives from its limited perspective, forms structures which thereafter determine what it further will and will not accept freely. From that point on, regardless of how the real world actually operates, this rational mind, following its self-imposed rules, tries to superimpose on the real world its own version of what must be.

This continues until at long last a beginner's mind cries out, "This is not right. What 'must be' is not happening. I have tried and tried to discover why this is so. I have stretched my imagination to the limit to preserve my belief in what 'must be.' The breaking point has come. Now I have no choice but to admit that the 'must' I have believed in does not come from the real world, but from my own head."

This narrative is not poetic hyperbole. It is a concise description of the major conclusion of the general theory of relativity and the means by which it was reached. The limited perspective is the perspective of our three-dimensional rationality and its view of one small part of the universe (the part into which we were born). The things that "must be" are the ideas of geometry (the rules governing straight lines, circles, triangles, etc.). The beginner's mind was Albert Einstein's. The long-held belief was that these rules govern, without exception, the entirety of the universe. What Einstein's beginner's mind realized was that this is so only in our minds.*

Einstein discovered that certain laws of geometry are valid only in limited regions of space. This makes them useful since our experience physically is limited to very small regions of space, like our solar system. However, as our experience expands, we encounter more and more difficulty in trying to superimpose these rules upon the entire expanse of the universe.

Einstein was the first person to see that the geometrical rules which apply to one small part of the universe as seen from a limited perspective (like ours) are not universal. This freed him to behold the universe in a way that no person had seen it before.

What he saw is the content of the general theory of relativity.

Einstein did not set out to prove anything about the nature of our minds. His interest was in physics. "Our new idea," he wrote, "is simple: to build a physics valid for all co-ordinate systems."[1] The fact that he *did* illustrate something of importance about the way that we structure our perceptions is indicative of an inevitable trend toward the merger of physics and psychology.

---

*The view presented here is not that geometry comes from the mind. There are many possible geometries (as Riemann and Lobachevsky showed before Einstein), but the actual geometry that we have is determined by the physics. For example, Euclid considered geometry to be closely related to experience (he defined congruence by moving triangles about in space) and he considered his parallel axiom to be not self-evident, i.e., not a product purely of the mind.

The view presented here is that idealizations abstracted from experience (like Euclidean geometry) form a rigid structure of such durability that, when subsequent sensory experience contradicts it, we question the validity of the sensory data rather than the validity of the idealized abstractions. Once such a set of idealized abstractions is erected (verified) in the mind, we thereafter superimpose it upon all subsequent actual and projected sense data (i.e., upon the entire universe as we picture it according to this set of abstractions), whether it fits or not.

How did Einstein get from a theory of physics to a revolutionary statement of geometry? How did that lead to a significant insight into our mental processes? The answer to these questions is one of the least known, but one of the most important and intriguing intellectual adventures recorded.

Einstein started with his special theory of relativity. As successful as it was, Einstein was not satisfied with it because it applied only to co-ordinate systems moving uniformly relative to each other. Is it possible, thought Einstein, to explain the same phenomenon as seen from two different frames of reference, one of them moving uniformly and the other of them moving non-uniformly, in such a way that there is a consistent explanation for the phenomenon in terms of both the uniformly moving frame of reference and the non-uniformly moving frame of reference. In other words, can we describe events which happen in a co-ordinate system which is moving non-uniformly in terms which are meaningful to an observer in a co-ordinate system which is moving uniformly, and the other way round? Can we create *one* physics that is valid for observers in both frames of reference?

Yes, discovered Einstein, it is possible for observers in the two different frames of reference to relate in a manner which is both meaningful in terms of their own state of motion and in terms of the other's state of motion. To illustrate this, he used another famous thought experiment.

Imagine an elevator in an extraordinarily tall building. The cable which supports the elevator has snapped, and the elevator is plummeting downward. Inside the elevator are several physicists. They are not aware that the cable has broken, and, since there are no windows, they cannot look outside.

The question is, what is the appraisal of this situation by the observers on the outside of the elevator (us) and by the observers on the inside of the elevator (the physicists)? Since this is an idealized experiment, we can disregard the effects of friction and the resistance of the air.

To us, the situation is apparent. The elevator is falling and soon it will strike the earth and all of its inhabitants will be dead. As the elevator falls, it accelerates according to Newton's law of gravity. The motion of the elevator is not uniform, but accelerated, because of the gravitational field of the earth.

We can predict many things that might happen inside the elevator. For example, if someone inside the elevator dropped a handkerchief, nothing would happen. It would appear to the inside observers to float where it was released because it would be accelerating toward the earth at the same rate as the elevator and the people inside of it. Nothing really would be floating, everything would be falling, but, since everything would be falling at the same rate, there would be no change in their relative positions.

To a generation of physicists born and brought up inside the elevator, however, things would appear quite differently. To them, dropped objects do not fall, they simply hang in midair. If someone gives a floating object a shove, off it goes in a straight line until it hits the side of the elevator. To the observers inside the elevator, there are no forces acting on any objects inside the elevator. In short, the observers inside the elevator would conclude that they are in an inertial co-ordinate system! The laws of mechanics are perfectly valid. Their experiments always produce results which agree exactly with theoretical predictions. An object at rest remains at rest. An object in motion remains in motion. Moving objects are deflected from their paths only by forces which are proportional to the amount of deflection. For every reaction there is an equal and opposite reaction. If we give a shove to a floating chair, it goes off in one direction, and we go off in the opposite direction with an equal momentum (although with a slower speed because of our greater mass).

The inside observers have a consistent explanation for the phenomena inside the elevator: They are in an inertial co-ordinate system, and they can prove it by the laws of mechanics.

The outside observers also have a consistent explanation for the phenomena inside the elevator: The elevator is falling in a gravitational field. Its passengers are unaware of this because, without being able to see outside the elevator, there is no way for them to detect it while they are falling. Their co-ordinate system is in accelerated motion, even though they believe that it is not moving at all.

The bridge between these two explanations is gravity.

The falling elevator is a pocket edition of an inertial co-ordinate system. A real inertial co-ordinate system is not limited in space or time. The elevator edition is limited in both. It is limited in space because a moving object inside the elevator will not move in a straight line forever, but only until it reaches one of the walls of the elevator. It is limited in time because sooner or later the elevator and its passengers are going to collide with the earth, ending their existence abruptly.

According to the special theory of relativity, moreover, it is significant that the elevator is limited in size because otherwise it would not appear to its inhabitants as an inertial co-ordinate system. For example, if the physicists inside the elevator simultaneously drop two baseballs, the baseballs float in the air exactly where they are released, and remain there. This, to the outside observer, is because they are falling parallel to each other. However, if the elevator were the size of Texas and the baseballs were as far apart when they were dropped as Texas is wide, the baseballs would not fall parallel to each other. They would *converge,* since each of them would be drawn by gravity to the center of the earth. The observers inside the elevator would notice that the baseballs, and any other floating objects in the elevator, move toward each other with the passage of time, as though there were a mutual attraction between them. This mutual attraction would appear as a "force" affecting the objects in the elevator, and the physicists inside hardly would conclude, under those circumstances, that they were in an inertial co-ordinate system.

In short, if it is small enough, *a co-ordinate system falling in a gravitational field is the equivalent of an inertial co-ordinate system.* This is Einstein's principle of equivalence. It is a telling piece of mental dexterity. Anything like an "inertial co-ordinate system" that can be "wiped out"[2] (Einstein's words) by the assumption of a gravitational field hardly deserves to be called absolute (as in "absolute motion," and "absolute nonmotion"). While the observers inside the elevator experience a lack of motion and the absence of gravity, the observers outside the elevator see a co-ordinate system (the elevator) accelerating through a gravitational field.

Now let us imagine a variation of this situation.

Assume that *we,* the outside observers, are in an inertial co-ordinate system. We already know what happens in inertial co-ordinate systems; the same things that happened in the falling elevator. There are no forces, including gravity, to affect us. Therefore, let us assume that we are comfortably floating. Objects at rest remain at rest, objects in motion continue in a straight line forever, and every action produces an equal and opposite reaction.

In our inertial co-ordinate system is an elevator. Someone has attached a rope to the elevator and is pulling it in the direction indicated.

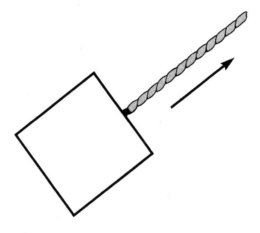

Since this is a thought experiment, it does not matter how this is done. The elevator is being pulled with a constant force, which means that it is in a state of constant acceleration in the direction of the arrow. How will observers outside the elevator and observers inside the elevator appraise this situation?

As we float outside the elevator, we experience that our frame of reference is absolutely at rest and that there is no gravity affecting it.

We see the elevator being pulled with a constant acceleration by the rope, and so we can predict certain things about it. Everything inside the elevator that is not attached quickly collides with the floor of the elevator. If someone in the elevator drops a handkerchief, the elevator floor rushes up to meet it. If someone in the elevator tries to jump off the floor, the floor, rushing upward, is instantly under his feet again. The floor of the elevator continually crashes into anything in its path as it accelerates upward.

Inside the elevator, however, the appraisal of the situation is quite different. To a generation of physicists born and brought up inside the elevator, talk of acceleration upward is fantasy (remember, the elevator has no windows). To them, their co-ordinate system is quite at rest. Objects fall downward to the floor because of a gravitational field, just as objects on the earth fall downward to the floor because of a gravitational field.

Both the observers inside the elevator and the observers outside the elevator have consistent explanations for the phenomena inside the elevator. We observers outside the elevator explain them by the accelerated motion of the elevator. The observers inside the elevator explain them by the presence of a gravitational field. *There is absolutely no way to determine which of us is right.*

"Wait a minute," we say, "suppose that we cut a small hole in one wall of the elevator and shine a light beam through it. If the elevator really were motionless, the light beam would strike the opposite wall of the elevator at a spot exactly opposite the hole. Since we can see that the elevator is accelerating upward, we know that the elevator wall will move upward slightly in the time it takes the light beam to cross the elevator. Therefore, the light beam will strike the far wall slightly below the spot just opposite the hole it entered through. In effect, it will seem to curve downward from the point of view of the people inside the elevator instead of traveling in a straight line. This should prove to them that their elevator is in motion."

"It does not prove anything of the sort," says Jim de Wit, who, of course, is inside the elevator. "The light beams in this elevator do not travel in straight lines. How could they? We are in a gravitational field. Light is energy, and energy has mass. Gravity attracts mass, and a light beam traveling through our elevator will be drawn downward by our gravitational field exactly like a baseball thrown horizontally at the speed of light."

There is no way that we can convince de Wit that his co-ordinate system is in a state of accelerated motion. Everything that we can say to prove this to him he dismisses (accounts for) as a result of his "gravitational field." *There is absolutely no way of distinguishing between uniform accelerated motion and a constant gravitational field.*

This is another expression of Einstein's principle of equivalence. In limited areas, *gravity is equivalent to acceleration.* We already saw that acceleration (falling) through a "gravitational field" is the equivalent of an inertial co-ordinate system. Now we see that a "gravitational field" is equivalent to accelerated motion. At last we are approaching a *general* theory of relativity, a theory valid for all frames of reference regardless of their states of motion.

The bridge which links the explanations of the observers inside of the elevator and the explanations of the observers outside of the elevator is gravity. The clue which indicated to Einstein that gravity was the key to his general theory was as old as physics itself.

There are two kinds of mass, which means that there are two ways of talking about it. The first is gravitational mass. The gravitational mass of an object, roughly speaking, is the weight of the object as measured on a balance scale. Something that weighs three times more than another object has three times more mass. Gravitational mass is the measure of how much force the gravity of the earth exerts on an object. Newton's laws describe the effects of this force, which vary with the distance of the mass from the earth. Although Newton's laws describe the effects of this force, they do not define it. This is the mystery of action-at-a-distance (page 24). How does the earth invisibly reach up and pull objects downward?

The second type of mass is inertial mass. Inertial mass is the measure of the resistance of an object to acceleration (or deceleration, which is negative acceleration). For example, it takes three times more force to move three railroad cars from a standstill to twenty miles per hour (positive acceleration) than it takes to move one railroad car from a standstill to twenty miles per hour (page 161). Similarly, once they are moving, it takes three times more force to stop three cars than it takes to stop the single car. This is because the inertial mass of the three railroad cars is three times more than the inertial mass of the single railroad car.

Inertial mass and gravitational mass are equal. This explains why a feather and a cannonball fall with equal velocity in a vacuum. The cannonball has hundreds of times more gravitational mass than the feather (it weighs more) but it also has hundreds of times more resistance to motion than the feather (its inertial mass). Its attraction to the earth is hundreds of times stronger than that of the feather, but then so is its inclination not to move. The result is that it accelerates downward at the same rate as the feather, although it seems that it should fall much faster.

The fact that inertial mass and gravitational mass are equal was known three hundred years ago, but physicists considered it a coincidence. No significance was attached to it until Einstein published his general theory of relativity.

The "coincidence" of the equivalence of gravitational mass and inertial mass was the "clew,"[3] to use Einstein's word, that led him to the principle of equivalence, which refers via the equivalence of gravitational mass and inertial mass to the equivalence of gravity and acceleration themselves. These are the things that he illustrated with his famous elevator examples.

The special theory of relativity deals with unaccelerated (uniform) motion.* If acceleration is neglected, the special theory of relativity applies. However, since gravity and acceleration are equivalent, this is the same as saying that the special theory of relativity is applica-

---

*The special theory deals with the unaccelerated (uniform) motion of *co-ordinate systems.* The special theory can be used to describe the accelerated (nonuniform) motion of *objects* as long as the co-ordinate system from which the object is being observed is itself in uniform motion.

ble whenever gravity is neglected. If the effects of gravity are to be considered, then we must use the general theory of relativity. In the physical world the effects of gravity can be neglected in (1) remote regions of space which are far from any centers of gravity (matter), and (2) in very small regions of space.

Why gravity can be ignored in very small regions of space leads to the most psychedelic aspect of all Einstein's theories. Gravity can be ignored in very small regions of space because, if the region is small enough, the mountainous terrain of space-time is not noticeable.*

The nature of the space-time continuum is like that of a hilly country-side. The hills are caused by pieces of matter (objects). The larger the piece of matter, the more it curves the space-time continuum. In remote regions of space far from any matter of significant size, the space-time continuum resembles a flat plain. A piece of matter the size of the earth causes quite a bump in the space-time continuum, and a piece of matter the size of a star causes a relative mountain.

As an object travels through the space-time continuum, it takes the easiest path between two points. The easiest path between two points in the space-time continuum is called a geodesic (geo dee' sic). A geodesic is not always a straight line owing to the nature of the terrain in which the object finds itself.

Suppose that we are in a balloon looking down on a mountain that has a bright beacon on the top of it. The mountain rises gradually out of the plain, and becomes more and more steep as its elevation increases, until, close to the top, it rises almost straight up. There are many villages surrounding the mountain, and there are footpaths connecting all of the villages with each other. As the paths approach the mountain, all of them begin to curve in one way or another, to avoid going unnecessarily far up the mountain.

---

*Some physicists think that general relativity will be useful on the microscale of high-energy physics (where the effects of gravity usually are ignored), e.g., strong fluctuations of the gravitational field have been detected at very short distances (10^{-14} cm).

Suppose that it is nighttime and that, looking down, we can see neither the mountain nor the footpaths. All that we can see is the beacon and the torches of the travelers below. As we watch, we notice that the torches deflect from a straight path when they approach the vicinity of the beacon. Some of them curve gently around the beacon in a graceful arc some distance away from it. Others approach the beacon more directly, but the closer they get to it, the more sharply they turn away from it.

From this, we probably would deduce that some force emanating from the beacon was repelling all attempts to approach it. For example, we might speculate that the beacon is extremely hot and painful to approach.

With the coming of daylight, however, we can see that the beacon is situated on the top of a large mountain and that it has nothing whatever to do with the movement of the torch-bearers. They simply followed the easiest paths available to them over the terrain between their points of origin and destination.

This masterful analogy was created by Bertrand Russell. In this case, the mountain is the sun, the travelers are the planets, asteroids, comets (and debris from the space program), the footpaths are their orbits, and the coming of daylight is the coming of Einstein's general theory of relativity.

The point is that the objects in the solar system move as they do not because of some mysterious force (gravity) exerted upon them at a distance by the sun, but because of the nature of the neighborhood through which they are traveling.

Arthur Eddington illustrated this same situation in another way. Suppose, he suggested, that we are in a boat looking down into clear water. We can see the sand on the bottom and the fishes swimming beneath us. As we watch, we notice that the fish seem to be repelled from a certain point. As they approach it, they swim either to the right or to the left of it, but never over it. From this we probably would deduce that there is a repellant force at that point which keeps the fish away.

However, if we should go into the water to get a closer look, we would see that an enormous sunfish has buried himself in the sand at that point, creating a sizable mound. As fish swimming along the bottom approach the mound, they follow the easiest path available to them, which is around it rather than over it. There is no "force" causing the fish to avoid that particular spot. If all had been known from the first, that spot was merely the top of a large mound which the fish found easier to swim around than to swim over.

The movement of the fish was determined not by a force emanating from the mysterious spot, but by the nature of the neighborhood through which they were passing. (Eddington's sunfish was called "Albert") (really). If we could see the geography (the geometry) of the space-time continuum, we would see that, similarly, it, and not "forces between objects," is the reason that planets move in the ways that they do.

It is not possible for us actually to see the geometry of the space-time continuum because it is four-dimensional and our sensory experience is limited to three dimensions. For that reason, it is not even possible to picture it.

For example, suppose that there existed a world of two-dimensional people. Such a world would look like a picture on a television or a movie screen. The people and the objects in a two-dimensional world would have height and width, but not depth. If these two-dimensional figures had a life and an intelligence of their own, their world would appear quite different to them than our world appears to us, for they could not experience the third dimension.

A straight line drawn between two of these people would appear to them as a wall. They would be able to walk around either end of it, but they would not be able to "step over" it, because their physical existence is limited to two dimensions. They cannot step off the screen into the third dimension. They would know what a circle is, but there is no way that they could know what a sphere is. In fact, a sphere would appear to them as a circle.

If they like to explore, they soon would discover that their world is flat and infinite. If two of them went off in opposite directions, they would never meet.

They also could create a simple geometry. Sooner or later they would generalize their experiences into abstractions to help them do and build the things that they want to do and build in their physical world. For example, they would discover that whenever three straight metal bars form a triangle, the angles of the triangle always total 180 degrees. Sooner or later, the more perceptive among them would substitute mental idealizations (straight lines) for the metal bars. That would allow them to arrive at the abstract conclusion that a triangle, which by definition is formed by three straight lines, always contains 180 degrees. To learn more about triangles, they no longer would need actually to construct them.

The geometry that such a two-dimensional people would create is the same geometry that we studied in school. It is called Euclidean geometry, in honor of the Greek, Euclid, whose thoughts on the subject were so thorough that no one expanded on them for nearly two thousand years. (The content of most high-school geometry books is about two millennia old.)

Now let us suppose that someone, unbeknownst to them, transported these two-dimensional people from their flat world onto the surface of an enormously large sphere. This means that instead of being perfectly flat, their physical world now would be somewhat curved. At first, no one would notice the difference. However, if their technology improved enough to allow them to begin to travel and to communicate over great distances, these people eventually would make a remarkable discovery. They would discover that their geometry could not be verified in their physical world.

For example, they would discover that if they surveyed a large enough triangle and measured the angles that form it, it would have more than 180 degrees! This is a simple phenomenon for *us* to picture. Imagine a triangle drawn on a globe. The apex (top) of the triangle is at the north pole. The two lines intersecting there form a right angle. The equator is the base of the triangle. Look what happens. Both sides of the triangle, upon intersecting the equator, also form right angles. According to Euclidean geometry, a triangle contains only two right angles (180 degrees), yet this triangle contains *three* right angles (270 degrees).

Remember that in our example, the two-dimensional people actually have surveyed a triangle on what they presumed was their flat world, measured the angles, and come up with 270 degrees. What a confusion. When the dust settles they would realize that there are only two possible explanations.

The first possible explanation is that the straight lines used to construct the triangle (like light beams) were not actually straight, although they seemed to be straight. This could account for the excessive number of degrees in the triangle. However, if this is the explanation that they choose to adopt, then they must create a "force" responsible for somehow distorting the straight lines (like "gravity"). The second possible explanation is that their abstract geometry does not apply to their real world. This is another way of saying that, impossible as it sounds, their universe is not Euclidean.

The idea that their physical reality is not Euclidean probably would sound so fantastic to them (especially if they had had no reason to question the reality of Euclidean geometry for two thousand years) that they probably would choose to look for forces responsible for distorting their straight lines.*

The problem is that, having chosen this course, they would be obligated to create a responsible force every time that their physical world failed to validate Euclidean geometry. Eventually the structure of these necessary forces would become so complex that it would be much simpler to forget them altogether and admit that their physical world does not follow the logically irrefutable rules of Euclidean geometry.

Our situation is parallel to that of the two-dimensional people

*Eddington expressed this concept most concisely: *"A field of force represents the discrepancy between the natural geometry of a co-ordinate system and the abstract geometry arbitrarily ascribed to it."* (Arthur Eddington, *The Mathematical Theory of Relativity*, Cambridge, England, Cambridge University Press, 1923, pp. 37–38. Italics in the original.)

who cannot perceive, but who can deduce that they are living in a three-dimensional world. We are a three-dimensional people who cannot perceive, but who can deduce that we are living in a four-dimensional universe.

For two thousand years we have assumed that the entire physical universe, like the geometry that the ancient Greeks created from their experience with this part of it, was Euclidean. That the geometry of Euclid is universally valid means that it can be verified anywhere in the physical world. That assumption was wrong. Einstein was the first person to see that the universe is not bound by the rules of Euclidean geometry, even though our minds tenaciously cling to the idea that it is.

Although we cannot perceive the four-dimensional space-time continuum directly, we can deduce from what we already know of the special theory of relativity that our universe is not Euclidean. Here is another of Einstein's thought experiments.

Imagine two concentric circles, one with a small radius and one with a very large radius. Both of them revolve around a common center as shown.

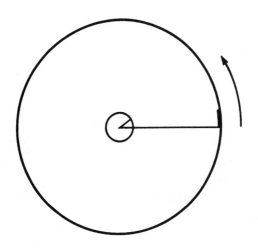

Imagine also that we, the observers, are watching these revolving circles from an inertial co-ordinate system. Being in an inertial co-ordinate system simply means that our frame of reference is at rest relative to everything, including the revolving circles. Drawn over the revolving circles are two identical concentric circles which are in our co-ordinate system. They are not revolving. They are the same size as the revolving circles and have the same common center, but they remain motionless. While we and our nonrevolving circles are motionless, we are in communication with an observer who is on the revolving circles. He actually is going around with them.

According to Euclidean geometry, the ratio of the radius to the circumference of all circles is the same. If we measure the radius and the circumference of the small circle, for example, the ratio of these two measurements will be the same as the ratio of the radius to the circumference of the large circle. The object of this thought experiment is to determine whether this is true or not for both the observers on the stationary circles (us) and the observer on the revolving circles. If the geometry of Euclid is valid throughout the physical universe, as it should be, we should discover that the ratio between the radius and the circumference of *all* the circles involved is identical.

Both we and the observer on the revolving circles will use the same ruler to do our measuring. "The same ruler" means that either we actually hand him the same ruler that we have used, or that we use rulers that have the same length when at rest in the same co-ordinate system.

We go first. Using our ruler, we measure the radius of our small circle, and then we measure the circumference of our small circle. Then we note the ratio between them. The next step is to measure the radius of our large circle and then the circumference of our large circle. Then we note the ratio between them. Yes, it is the same ratio that we found between the radius and the circumference of our small circle. We have proved that Euclidean geometry *is* valid in our co-ordinate system, which is an inertial co-ordinate system.

Now we hand the ruler to the observer on the revolving circles as he passes by us. Using this ruler he first measures the radius of his small circle and finds that it is the same as ours, since our circles are drawn directly over his circles. Next he measures the circumference of his small circle. Remember that motion causes rulers to contract in the direction that they are moving. However, since the radius of the small circle is so short, the velocity of the ruler when it is placed on the circumference of the small circle is not fast enough to make the effect of relativistic contraction noticeable. Therefore, the observer on the revolving circles measures the circumference of his small circle and finds it to be the same as the circumference of our small circle. Naturally, the ratio between them also is the same. So far so good. The ratios between the radius and the circumference of three circles have been determined (our small circle, our large circle, and his small circle) and they are all identical. This is exactly what should happen according to high-school geometry books across the country. Only one more circle to go.

The observer on the revolving circles measures the radius of his large circle and finds it to be the same length as the radius of our large circle. Now he comes to the last measurement, the circumference of his large circle. However, as soon as he puts his ruler into position to make a measurement on the circumference of the large revolving circle, his ruler contracts! Because the radius of his large circle is much larger than the radius of his small circle, the velocity of the circumference of the large revolving circle is considerably faster than the velocity of the circumference of the small revolving circle.

Since the ruler must be aligned in the direction that the circumference is moving, it becomes shorter. When the revolving observer uses this ruler to measure the circumference of the large revolving circle, he finds that it is larger than the circumference of our large circle. This is because his ruler is shorter. (Contraction also affected his ruler when he measured the radius of his large circle, but since it then was placed perpendicular to the direction of motion, it became skinnier, nor shorter).

This means that the ratio of the radius to the circumference of the small revolving circle is not the same as the ratio of the radius to the circumference of the large revolving circle. According to Euclidean geometry, this is not possible, but there it is.

If we want to be old-fashioned about it (before-Einstein) we can say that this situation is nothing unusual. By definition, the laws of mechanics and the geometry of Euclid are valid only in inertial systems (that is what makes them inertial systems). We simply don't consider co-ordinate systems which are not inertial. (This was really the position of physicists before Albert Einstein.) This is exactly what seemed wrong to Einstein. His idea was to create a physics valid for *all* co-ordinate systems, since the universe abounds with the non-inertial as well as the inertial kind.

If we are to create such a universally valid physics, a general physics, then we must treat both the observers in the stationary (inertial) system and the observer on the revolving circles (a non-inertial system) with equal seriousness. The person on the revolving circles has as much right to relate the physical world to his frame of reference as we have to relate it to ours. True, the laws of mechanics as well as the geometry of Euclid are not valid in his frame of reference, but every deviation from them can be explained in terms of a gravitational field which affects his frame of reference.

This is what Einstein's theory allows us to do. It allows us to express the laws of physics in such a way that they are independent of specific space-time co-ordinates. Space and time co-ordinates (measurements) vary from one frame of reference to another, depending upon the state of motion of the frame of reference. The general theory of relativity allows us to universalize the laws of physics and to apply them to all frames of reference.

"Wait a minute," we say, "how can anyone measure distance or navigate in a co-ordinate system like the one on the revolving circles? The length of a ruler varies from place to place in such a system. The farther we go from the center, the faster the velocity of the ruler, and the more it contracts. This doesn't happen in an inertial co-ordinate system, which, in effect, is a system that is at rest. Because there is no change of velocity throughout an inertial co-ordinate system, rulers do not change length.

"This allows us to organize inertial systems like a city, block by block. Since rulers do not change length in inertial systems, all the blocks that are laid out with the same ruler will be the same length. No matter where we travel, we know that ten blocks is twice the distance of five blocks.

"In a non-inertial system the velocity of the system varies from place to place. This means that the length of a ruler varies from place to place. If we used the same ruler to lay out all the city blocks in a non-inertial co-ordinate system, some of them would be larger than others depending upon where they were located."

"What is wrong with that," asks Jim de Wit, "as long as we still can determine our position in the co-ordinate system? Imagine a sheet of india rubber on which we have drawn a grid so that it looks like a piece of graph paper (first drawing, next page). This is a co-ordinate system. Assuming that we are at the lower left corner (we can start anywhere) let us say that a party Saturday night is being held at the intersection marked 'Party.' To get there we have to go two squares to the right and two squares up.

"Now suppose that we stretch the sheet of rubber so that it looks like the second drawing.

"The same directions (two squares right and two squares up) still bring us to the party. The only difference is that unless we are familiar with this part of the co-ordinate system, we cannot calculate the distance that we have to travel as easily as we could if all of the squares were the same size."

According to the general theory of relativity, gravity, which is the equivalent of acceleration, is what distorts the space-time continuum in a manner analogous to our stretching the sheet of rubber. Where the effects of gravity can be neglected, the space-time continuum is like the sheet of rubber before we stretched it. All of the lines are straight lines and all of the clocks are synchronized. In other words, the undistorted sheet of rubber is analogous to the space-time contin-

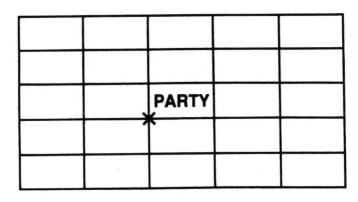

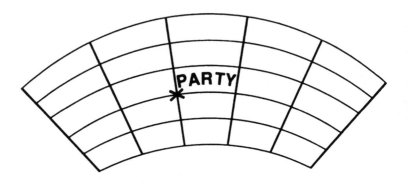

uum of an inertial co-ordinate system and the special theory of relativity applies.

However, in the universe at large gravity cannot be neglected. Wherever there is a piece of matter, it warps the space-time continuum. The larger the piece of matter, the more pronounced the warp.

In the example of the revolving circles, the variation of velocity in different parts of the co-ordinate system caused the ruler to change size. With that in mind, remember that *acceleration (change in velocity) is the equivalent of gravity*. Therefore, changes in the strength of a gravitational field will produce the same contractions of the ruler as changes in velocity. "Acceleration" and "gravity" are two ways of saying the same thing. That means that if a ruler is subjected to gravitational fields of different strength, it changes length.

Of course, it is impossible to travel through our solar system, much less our galaxy, without encountering gravitational fields of varying intensity, which would cause any maps that we somehow could produce to look distorted like the stretched piece of india rubber. The terrain of the space-time continuum in which our earth moves is like a hilly countryside with a mountain (the sun) dominating the geography.

According to Newton, the earth wants to continue forever in a straight line, but forever is deflected from its inclination by the gravitational force of the sun. A balance of the two keeps the earth in orbit around the sun. According to Einstein, the earth's orbit is simply the easiest path for the earth to take as it moves through the space-time continuum, warped as it is in this neighborhood by the sun.

Imagine how complex is the geography of the space-time continuum which is our universe with its solar systems, star systems, galaxies, and galaxy clusters, each of them causing major and minor bumps, curves, hills, valleys, and mountains in the four-dimensional space-time continuum.

Would it be possible to navigate under such circumstances?

Yes. Although it is a crude example, sailors navigate under somewhat analogous circumstances. We cover the earth with squares which are formed by lines of latitude and longitude. The size of these squares varies depending upon where they are located. The closer they are to the equator, the larger they are. (If this is unclear, look at a globe). Nonetheless, we still can locate physical points on the surface of the earth by designating the intersection of a line of latitude and a line of longitude. Knowing the number of squares between us and where we want to sail does not give us the distance to our destination because the squares may vary in size. However, if we know the nature of our terrain (a globe) we can calculate distances on it (using spherical trigonometry).

Similarly, once we know the properties of an area of the space-time continuum (by exploring it) we can determine not only the position of, but also the distance (interval) between two events in the space-time continuum.* The mathematical structure of the general theory of relativity, which Einstein created over a period of ten years, permits us to do just that.

The equations of the general theory of relativity are structural formulas. They describe the structure of changing gravitational fields. (Newton's formula describes a situation between two objects at a given time. Einstein's formulas relate a situation here and now to a situation in the immediate vicinity a little later.) By feeding the results of actual observations into these equations, they give us a picture of the space-time continuum in the neighborhood of our observations. In other words, they reveal the geometry of space-time in that area. Once we know that, our situation is roughly analogous to that of a sailor who knows that the earth is round and also knows spherical trigonometry.†

We have said, up to now, that matter distorts, or causes a curvature of, the space-time continuum in its vicinity. According to Einstein's ultimate vision, which he never "proved" (demonstrated mathematically), a piece of matter *is* a curvature of the space-time continuum! In other words, according to Einstein's ultimate vision, there are no such things as "gravitational fields" and "masses." They are only mental creations. No such things exist in the real world. There is no such thing as "gravity"—gravity is the equivalent of acceleration, which is motion. There is no such thing as "matter"—matter is a curvature of the space-time continuum. There is not even such a thing as "energy"—energy equals mass and mass is space-time curvature.

What we considered to be a planet with its own gravitational field moving around the sun in an orbit created by the gravitational attraction (force) of the sun is actually a pronounced curvature of the

---

*This distance, of course, is "invariant," i.e., the same for all co-ordinate systems (page 171). The invariance is the absolute objective aspect of Einstein's theory that complements the subjective arbitrary choice of co-ordinate system.

†The space-time continuum is not only curved, it also has topological properties, i.e., it can be connected in crazy ways, e.g., like a donut ◎. It also can twist (i.e., torsion).

space-time continuum finding its easiest path through the space-time continuum in the vicinity of a very pronounced curvature of the space-time continuum.

There is nothing but space-time and motion and they, in effect, are the same thing. Here is an exquisite presentation, in completely western terms, of the most fundamental aspect of Taoist and Buddhist philosophies.

Physics is the study of physical reality. If a theory does not relate to the physical world, it may be pure mathematics, poetry, or blank verse, but it is not physics. The question is, does Einstein's fantastic theory really work?

The answer is a slightly tentative, but generally accepted "Yes." Most physicists agree that the general theory of relativity is a valid way of viewing large-scale phenomena, and at the same time, most physicists still are eager to see more evidence to confirm (or challenge) this position.

Since the general theory of relativity deals with vast expanses of the universe, its proof (or usefulness, not of "truth"—the watch is still unopenable) cannot come from observations of phenomena limited to the earth. For this reason, its verifications come from astronomy.

Thus far, the general theory of relativity has been verified in four ways. The first three ways are straightforward and convincing. The last way, if early observations are correct, may be more fantastic than the theory itself.

The first verification of the general theory of relativity came as an unexpected benefit to astronomers. Newton's law of gravity purported to describe the orbits of the planets around the sun, and it did—all of them except Mercury. Mercury orbits the sun in such a way that some parts of its orbit bring it closer to the sun than others. The part of Mercury's orbit closest to the sun is called its perihelion. The first verification of Einstein's general theory of relativity turned out to be the long-sought explanation of the problem of Mercury's perihelion.

The problem with Mercury's perihelion—in fact, with Mercury's entire orbit—is that it moves. Instead of continuously retracing its path around the sun relative to a co-ordinate system attached to the sun, Mercury's orbit itself revolves around the sun. The rate of revolution is extremely slow (it completes one revolution around the sun every three million years). This still was enough to puzzle astronomers. Prior to Einstein, this precession in Mercury's orbit had been attributed to an undiscovered planet in our solar system. By the time Einstein published his general theory of relativity, the search for this mysterious planet was well underway.

Einstein created his general theory of relativity without special attention to the perihelion of Mercury. However, when the general theory of relativity was applied to this problem, it showed that Mercury moves precisely as Mercury has to move through the space-time continuum in that vicinity of the sun! The other planets do not move significantly in this way because they are farther away from the sun's gravity. Score one for the general theory.

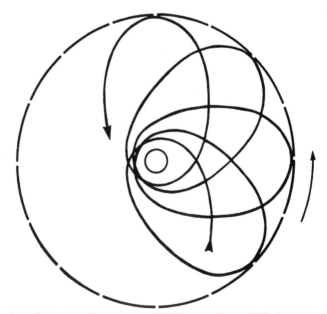

The second verification of the general theory of relativity was the fulfillment of a prediction specifically made by Einstein. Einstein predicted that light beams are bent by gravitational fields. He also predicted exactly how much they are bent, and he suggested an experiment to test this prediction. Einstein suggested that astronomers measure the deflection of starlight by the gravitational field of the sun.

According to Einstein, the presence of the sun between a group of visible stars and the earth will cause an apparent change in the position of the stars because light coming from them will be bent by the gravitational field of the sun. In order to perform this experiment, it is necessary to photograph a group of stars at night, noting their positions relative to each other and other stars in their periphery, and then to photograph the same group during the day when the sun is between them and us. Of course, stars only can be photographed in the daytime during a total eclipse of the sun by the moon.

Astronomers consulted their star charts and discovered that May 29 is the ideal day for such an undertaking. This is because the sun, in its apparent journey across a varied stellar background, is in front of an exceptionally rich grouping of bright stars on that date. By incredible coincidence, a total eclipse of the sun occurred on May 29, 1919, only four years after the general theory was published. Preparations were made to use this event to test Einstein's new theory.

Light signals from a star are bent in the neighborhood of the sun. Because we assume that starlight travels in a straight line, we assume that the star is in a position other than it actually is.

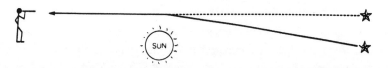

Although light was supposed to travel in a straight line in a vacuum, a certain amount of bending already was theorized before Einstein's general theory of relativity. Newton's law of gravity was used to calculate this bending, even though it could not explain it. Einstein's theory predicted roughly *twice* the deflection that Newton's law predicted, and, in addition, it supplied an explanation for it. Physicists and astronomers alike eagerly awaited the outcome of this confrontation between the new theory and the old.

The 1919 eclipse was photographed by two different expeditions sent to two different parts of the world. These expeditions also took photographs of the same stellar background at times when the sun was not in the area. The results of both expeditions vindicated Einstein's calculations, not Newton's. Since 1919, the same verdict has been reached again and again during other eclipses. All of them confirm Einstein's predictions. Score two for the general theory.

The third verification of the general theory of relativity is called gravitational redshift. Remember that gravity (because it is the equivalent of acceleration) not only causes rulers to contract, but it also causes clocks to run more slowly.

A clock is anything that repeats itself periodically. An atom is a type of clock. It vibrates at a certain frequency. When a substance, like sodium, is made to glow, the wavelength of the light that it emits can be measured accurately. This wavelength tells us exactly the frequency of the vibrations of the atoms that comprise the substance. If the frequency should vary, the wavelength also will vary.

If we want to compare the rhythm of a clock here on the earth with the rhythm of a clock that is influenced by an intense gravitational field, like that of the sun, we do not need to send a clock to the surface of the sun. The clocks already are in place.

Einstein predicted that any periodic process that takes place in an atom on the sun, where the gravity is very intense, must take place at a slightly slower rate than it does here on the earth. To test this prediction, all we need do is compare the wavelength of the radiation of a given element as it is found in sunlight and as it is found here on earth in the laboratory. This has been done many times. In each case, the wavelength measured from the sunlight was found to be longer than its laboratory counterpart. A longer wavelength means a lower (slower) frequency. Sodium atoms, for example, vibrate more slowly under the influence of the sun's strong gravitational field than they do on the earth. So do all the atoms.

This phenomenon is called gravitational redshift because the wavelengths involved appear to be shifted slightly toward the red end of the visible light spectrum where the wavelengths are the longest. Score three for the general theory.

Mercury's moving perihelion, starlight deflection, and gravitational redshift are all observable phenomena. Now we come to an area where theory is still predominant and observation is minimal. Nonetheless, it is an area that is by far the most exciting and perhaps the most stimulating in the entire history of science. The fourth verification of the general theory of relativity appears to be the phenomenon of the black hole.

In 1958, David Finkelstein published a paper in which he theorized, on the basis of Einstein's general theory of relativity, a phenomenon that he called a "one-way membrane."[4] Finkelstein showed that under certain conditions involving an extremely dense gravitational field, an invisible threshold can occur into which light and physical objects can enter, but from which they never again can escape.*

The following year, a young graduate student at the University of London heard Finkelstein, who was speaking there as a guest lecturer, explain his one-way membrane. The idea caught his attention and then his imagination. The young student was Roger Penrose. Expanding on Finkelstein's discovery, he developed it into the modern theory of the "Black Hole."†

A black hole is an area of space which appears absolutely black

---

*This phenomenon was theorized by Pierre-Simon La Place in 1795 using Newtonian physics. Finkelstein was the first physicist to formulate it from the modern point of view, i.e., relativity theory. This modern formulation triggered the current theories on the black hole.

†The very first modern paper on black holes was done by J. R. Oppenheimer and S. Snyder in 1939. The current theories of the black hole, i.e., black hole singularities which are beyond space-time, were developed independently by R. Penrose and S. W. Hawking.

because the gravitation there is so intense that not even light can escape into the surrounding areas.* Gravitation is negligible on the laboratory level, but quite important when bodies of large mass are concerned. Therefore, the exploration of black holes naturally became a joint venture of physicists and astronomers.

Astronomers speculated that a black hole may be one of several possible products of stellar evolution. Stars do not burn indefinitely. They evolve through a life cycle which begins with hydrogen gas and sometimes ends with a very dense, burned-out, rotating mass. The exact end product of this process depends upon the size of the star undergoing it. According to one theory, stars which are about three times the size of our sun or larger end up as black holes. The remains of such stars are unimaginably dense. They may be only a few miles in diameter and yet contain the entire mass of a star three times larger than the sun. Such a dense mass produces a gravitational field strong enough to pull everything in its vicinity into it, while at the same time allowing nothing, not even light, to escape from it.

Surrounding this remainder of a star is an "event horizon." An event horizon is created by the enormous gravitational field of the burned-out star. It functions precisely like Finkelstein's one-way membrane. Anything within the gravitational field of this mass quickly is pulled toward it, and once past the event horizon, never can return. It is the event horizon which constitutes the essential feature of the black hole. What happens to an object that passes through an event horizon is even more fantastic than the wildest (currently) science fiction.

If the black hole is not rotating, the object will be pulled directly to the center of the black hole to a point called the singularity. There it literally will be squeezed out of existence, or as physicists say, to zero volume. At the black hole singularity all of the laws of physics break down completely, and even space and time disappear. It is spec-

---

*To a first approximation. Physicists currently theorize that black holes actually shine due to photons and other particles quantum-tunneling out of the one-way membranes.

ulated that everything which is sucked into a black hole is spilled out again on "the other side"—the "other side" being another universe!

If the black hole is rotating, an object that is sucked into the event horizon could miss the black hole singularity (which is shaped like a "ring" in a rotating black hole) and emerge into another time and another place in this universe (through "wormholes"), or into another universe (through "Einstein-Rosen bridges"). In this way, rotating black holes may be the ultimate time machines.

Although black holes are almost invisible, we can search for observable phenomena that may be characteristic of them. The first of these is a large amount of electromagnetic radiation. A black hole continuously attracts hydrogen atoms, cosmic particles, and everything else to it. As these particles and objects are drawn to the black hole, they steadily accelerate through its gravitational field until they approach the velocity of light itself. This causes tremendous amounts of electromagnetic radiation. (Any accelerating charged particle creates electromagnetic radiation.)

The second observable characteristic of an invisible black hole is its effect on a nearby visible star. If a visible star can be found which moves as though it were revolving around an invisible star (i.e., as though it were half of a binary star system), we might speculate that it actually *is* revolving around an invisible star, and that its invisible partner is a black hole.

The search for black holes consequently became the search for these two phenomena. In 1970, the satellite Uhuru located both of them in one area. It pinpointed a high-energy x-ray source in the constellation Cygnus which emits a million times more energy than the sun. This high-energy source of electromagnetic radiation, which came to be known as Cygnus X-1, is very close to a visible blue-hot supergiant star. Scientists now believe that this blue supergiant forms a binary system with the black hole, Cygnus X-1.

As the visible star and the invisible black hole orbit each other, the blue supergiant literally is being sucked into the black hole. As material is torn away from its surface, it plunges into the black hole at tremendous speed, emitting x-rays. Incredible as Cygnus X-1 is, more than one hundred similar objects have been detected within our own Milky Way galaxy since its discovery. Although black holes stretch our imagination to the limit, the evidence is mounting that they actually do exist.

For example, if black holes are as we have speculated them to be, whatever disappears in them reappears somewhere. Is it possible, therefore, that there are black holes in other universes which are sucking matter from those universes into our universe? This is a seriously considered possibility. There are objects in our universe that appear to be the reverse of black holes. They are called white holes (of course). These objects are quasi-stellar radio sources, or quasars for short.

Quasars are extraordinarily intense energy sources. Most of them are only several times the diameter of our solar system, yet they emit more energy than an entire galaxy of over 150 billion stars! Some astronomers believe that quasars are the most distant objects ever detected, yet their incredible brightness allows us to see them clearly.

The relationship between black holes and quasars is purely speculation, but the speculation is mind-boggling. For example, some physicists speculate that black holes swallow up matter from one universe and pump it either into another universe or into another part and time of the same universe. The "output" side of a black hole, according to this hypothesis, is a quasar. If this speculation is correct, then our universe is being sucked into its many black holes, only to reappear in other universes, while other universes are being pumped into our own universe, which is being sucked through black holes and into other universes again. The process goes on and on, feeding on itself, another beginningless, endless, endless, beginningless dance.

One of the most profound by-products of the general theory of relativity is the discovery that gravitational "force," which we had so long taken to be a real and independently existing thing, is actually our mental creation. There is no such thing in the real world. The planets do not orbit the sun because the sun exerts an invisible gravitational force on them, they follow the paths that they do because those paths are the easiest ways for them to traverse the terrain of the space-time continuum in which they find themselves.

The same is true for "nonsense." It is a mental creation. There is no such thing in the real world. From one frame of reference black holes and event horizons make sense. From another frame of reference absolute nonmotion makes sense. Neither is "nonsense" except as seen from another point of view.

We call something nonsense if it does not agree with the rational edifices that we carefully have constructed. However, there is nothing intrinsically valuable about these edifices. In fact, they themselves often are replaced by more useful ones. When that happens, what was nonsensical from an old frame of reference can make sense from a new frame of reference, and the other way round. Like measurements of space and time, the concept of nonsense (itself a type of measurement) is relative, and we always can be sure when we use it that from some frame of reference it applies to us.

摇

Part One

I CLUTCH
MY IDEAS

1

🌀

The Particle Zoo

*The fourth translation of Wu Li is "I Clutch My Ideas." This is appro-*priate to a book on physics since the history of science in general often has been the story of scientists vigorously fighting an onslaught of new ideas. This is because it is difficult to relinquish the sense of security that comes from a long and rewarding acquaintance with a particular world view.

The value of a physical theory depends upon its usefulness. In this sense the history of physical theories might be said to resemble the history of individual personality traits. Most of us respond to our environment with a collection of automatic responses that once brought desirable results, usually in childhood. Unfortunately, if the environment that produced these responses changes (we grow up) and the responses themselves do not adapt, they become counterproductive. Showing anger, becoming depressed, flattering, crying, and bullying behavior are response patterns appropriate to times often long past. These patterns change only when we are forced to realize that they are no longer productive. Even then change is often painful and slow. The same is true of scientific theories.

Not one person, except Copernicus, wanted to accept the Copernican idea that the earth revolves around the sun. Goethe wrote about the Copernican revolution:

> Perhaps a greater demand has never been laid upon mankind; for by this admission [that the earth is not the center of the universe], how much else did not collapse in dust and smoke: a second paradise, a world of innocence, poetry, and piety, the witness of the senses, the convictions of a poetic and religious faith; no wonder that men had no stomach for all this, that they ranged themselves in every way against such a doctrine . . .[1]

Not one physicist, not even Planck himself, wanted to accept the implications of Planck's discovery, for to do so threatened a scientific structure (Newtonian physics) over three hundred years old. Heisenberg wrote about the quantum revolution:

> . . . when new groups of phenomena compel changes in the pattern of thought . . . even the most eminent of physicists find immense difficulties. For the demand for change in the thought pattern may engender the feeling that the ground is to be pulled from under one's feet. . . . I believe that the difficulties at this point can hardly be overestimated. Once one has experienced the desperation with which clever and conciliatory men of science react to the demand for a change in the thought pattern, one can only be amazed that such revolutions in science have actually been possible at all.[2]

Scientific revolutions are forced upon us by the discovery of phenomena that are not comprehensible in terms of the old theories. Old theories die hard. Much more is at stake than the theories themselves. To give up our privileged position at the center of the universe, as Copernicus asked, was an enormous psychological task. To accept that nature is fundamentally irrational (governed by chance), which is the essential statement of quantum mechanics, is a powerful blow to the intellect. Nonetheless, as new theories demonstrate superior utility, their adversaries, however reluctantly, have little choice but to accept them. In so doing, they also must grant a measure of recognition to the world views that accompany them.

Today, particle accelerators, bubble chambers, and computer printouts are giving birth to another world view. This world view is as different from the world view at the beginning of this century as the Copernican world view was from its predecessors. It calls upon us to relinquish many of our closely clutched ideas.

In this world view there is no substance.

The most common question that we can ask about an object is, "What is it made of?" That question, however, "What is it made of?" is based upon an artificial mental structure that is much like a hall of mirrors. If we stand directly between two mirrors and look into one, we see our reflection, and, just behind ourselves, we see a crowd of "us"s, each looking at the back of the head in front of it, stretching backward as far as we can see. These reflections, all of them, are illusions. The only real thing in the whole setting is *us (we)*.

This situation is very similar to what happens whenever we ask of something, "What is it made of?" The answer to such a question is always another something to which we can apply the same question.

Suppose, for example, that we ask of an ordinary toothpick, "What is it made of?" The answer, of course, is "wood." However, the question itself has taken us into a hall of mirrors because now we can ask about the wood, "What is it made of?" Closer examination reveals that wood is made of fibers, but what the fibers are made of is another question, and so on.

Like a pair of parallel mirrors, reflecting reflections, gives the illusion of an unending progression to nowhere, the idea that a thing can be different from what it is made of creates an infinite progression of answers, leaving us forever frustrated in an unending search. No matter what something—anything—is "made of," we have created an illusion which forces us to ask, "Yes, but what is *that* made of?"

Physicists are people who have pursued tenaciously this endless series of questions. What they have found is startling. Wood fibers, to continue the example, are actually patterns of cells. Cells, under magnification, are revealed to be patterns of molecules. Molecules, under higher magnification, are discovered to be patterns of atoms, and, lastly, atoms have turned out to be patterns of subatomic particles. In other words, "matter" is actually a series of *patterns out of focus*. The search for the ultimate stuff of the universe ends with the discovery that there *isn't any*.

If there is any ultimate stuff of the universe, it is pure energy, but subatomic particles are not "made of" energy, they *are* energy. This is what Einstein theorized in 1905. Subatomic interactions, therefore, are interactions of energy with energy. At the subatomic level there is no longer a clear distinction between what is and what happens, between the actor and the action. At the subatomic level the dancer and the dance are one.

According to particle physics, the world is fundamentally dancing energy; energy that is everywhere and incessantly assuming first this form and then that. What we have been calling matter (particles) constantly is being created, annihilated, and created again. This happens as particles interact and it also happens, literally, out of nowhere.

Where there was "nothing" there suddenly is "something," and then the something is gone again, often changing into something else before vanishing. In particle physics there is no distinction between empty, as in "empty space," and not-empty, or between something and not-something. The world of particle physics is a world of sparkling energy forever dancing with itself in the form of its particles as they twinkle in and out of existence, collide, transmute, and disappear again.

The world view of particle physics is a picture of *chaos beneath order*. At the fundamental level is a confusion of continual creation, annihilation and transformation. Above this confusion, limiting the forms that it can take, are a set of conservation laws (page 176). They do not specify what must happen, as ordinary laws of physics do, rather they specify what can*not* happen. They are permissive laws. At the subatomic level, absolutely everything that is not forbidden by the conservation laws actually happens. (Quantum theory describes the probabilities of the possibilities permitted by the conservation laws).

The old world view was a picture of order beneath chaos. It

assumed that beneath the prolific confusion of detail that constitutes our daily experience lie systematic and rational laws which relate them one and all. This was Newton's great insight: The same laws which govern falling apples govern the motion of planets. There is still, of course, much truth in this, but the world view of particle physics is essentially the opposite.

The world view of particle physics is that of a world without "stuff," where what is = what happens, and where an unending tumultuous dance of creation, annihilation, and transformation runs unabated within a framework of conservation laws and probability.

High-energy particle physics is the study of subatomic particles. It usually is shortened to "particle physics." Quantum theory and relativity are the theoretical tools of particle physics. The hardware of particle physics is housed in unimaginably expensive facilities which couple particle accelerators and computers.

The original purpose of particle physics was to discover the ultimate building blocks of the universe. This was to be accomplished by breaking matter into smaller and smaller pieces, eventually arriving at the smallest pieces possible. The experimental results of particle physics, however, have not been so simple Today most particle physicists are engaged in making sense out of their copious findings.*

In principle, particle physics hardly could be simpler. Physicists send subatomic particles smashing into each other as hard as they can. They use one particle to shatter another particle so that they can see what the remains are made of. The particle that does the smashing is called the projectile and the particle that gets smashed is called the target. The most advanced (and expensive) particle accelerators send

*The present state of high-energy theory is similar to Ptolemaic astronomy before its collapse under the pressure of the new Copernican world view. The discovery of new particles and new quantum numbers, e.g., charm (to be discussed later) is analogous to the addition of epicycles piled on an already unwieldy theoretical structure.

both the projectile and the target particles flying toward a common collision point.

The collision point usually is located inside a device called a bubble chamber. As charged particles move through a bubble chamber, they leave trails similar to the vapor trails that jetliners leave in the atmosphere. The bubble chamber is located inside a magnetic field. This causes particles with a positive charge to curve in one direction and particles with a negative charge to curve in the opposite direction. The mass of the particle can be determined by the tightness of the curve that the particle makes (lighter particles curve more than heavier particles with the same velocity and charge). A computer-triggered camera makes a photograph every time a particle enters the bubble chamber.

This elaborate arrangement is necessary because most particles live much less than a millionth of a second and are too small to be observed directly.* In general, everything a particle physicist knows about subatomic particles, he deduces from his theories and from photographs of the tracks that particles leave in a bubble chamber.†

Bubble chamber photographs, thousands and thousands of them, show clearly the frustrating situation which early particle physicists encountered in their search for "elementary" particles. When the projectile strikes the target, both particles are destroyed at the point of impact. In their place, however, are created *new* particles, all of which are as "elementary" as the original particles and often as massive as the original particles!

The schematic diagram on the next page shows a typical particle interaction. A particle called a negative pi meson (π^-) collides with a proton (p). Both the pi meson and the proton are destroyed and in their place are created two new particles, a neutral K meson (K^0) and a lambda particle (Λ). Both of these particles decay spontaneously

---

*The dark-adapted eye can detect single photons. All of the other subatomic particles must be detected indirectly.

†In addition to bubble-chamber physics there is emulsion (photographic plate) physics, counter physics, etc. However, the bubble chamber is probably the most commonly used detection device in particle physics.

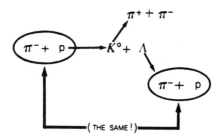

(no collision necessary) into two additional particles, leaving four new particles. Of these four particles, two of them are the same particles that we started with! It is as though, wrote Finkelstein, we fling two clocks together, they shatter, and out of them come flying not gears and springs but more clocks, some of them as large as the originals.

How can this happen? The answer is partly given by Einstein's special theory of relativity. *The new particles are created from the kinetic energy (energy of motion) of the projectile particle* in addition to the mass of the projectile particle and the mass of the target particle. The faster the projectile particle is traveling, the more kinetic energy is available to create new particles at the point of impact. For this reason, governments have spent more and more money to construct larger and larger particle accelerators which can push projectile particles to higher and higher velocities. If both the projectile particle and the target particle are accelerated to the point of impact, so much the more kinetic energy is available to create new particles to study.

Every subatomic interaction consists of the annihilation of the original particles and the creation of new subatomic particles. The subatomic world is a continual dance of creation and annihilation, of mass changing to energy and energy changing to mass.* Transient forms

---

*The mass/energy dualism of our ordinary conceptualizations does not exist in the formalism of relativity or quantum theory. According to Einstein's $E = mc^2$, mass does not change into energy or vice versa: Energy *is* mass. Wherever energy, E, is present, mass, m, is present and the amount of mass, m, is given by $E = mc^2$. The total amount of energy, E, is conserved, and hence the total amount of mass, m, also is conserved. This mass, m, is defined by the fact that it is a source of the gravitational field.

sparkle in and out of existence creating a never-ending, forever-newly-created reality.

Mystics from both the East and the West who claim to have beheld "the face of God" speak in terms so similar to these that any psychologist who professes an interest in altered states of awareness scarcely can ignore this obvious bridge between the disciplines of physics and psychology.

The first question of particle physics is, "What collides?"

According to quantum mechanics, a subatomic particle is not a particle like a particle of dust. Rather, subatomic particles are "tendencies to exist' (page 35) and "correlations between macroscopic observables" (page 78). They have no objective existence. That means that we cannot assume, if we are to use quantum theory, that particles have an existence apart from their interactions with a measuring device (page 105). As Heisenberg wrote:

> In the light of the quantum theory . . . elementary particles are
> no longer real in the same sense as objects of daily life, trees or
> stones . . .[3]

When an electron, for example, passes through a photographic plate it leaves a visible "track" behind it. This "track," under close examination, is actually a series of dots. Each dot is a grain of silver formed by the electron's interaction with atoms in the photographic plate. When we look at the track under a microscope, it looks something like this.

Ordinarily we would assume that one and the same electron, like a little baseball, went streaking through the photographic plate and left this trail of silver grains behind it. This is a mistake. Quantum mechanics tells us the same thing that Tantric Buddhists have been saying for a millennium. *The connection between the dots (the "moving*

object") is a product of our minds and it is not really there. In rigorous quantum mechanical terms, the moving object—the particle with an independent existence—is an unprovable assumption.

"According to our customary way of reasoning," wrote David Bohm, a professor of physics at Birkbeck College, University of London,

> we could suppose that the track of grains of silver indicates that a real electron moves continuously through space in a path somewhere near these grains, and by interaction caused the formation of the grains. But according to the usual interpretation of the quantum theory, it would be incorrect to suppose that this really happened. All that we can say is that certain grains appeared, but we must not try to imagine that these grains were produced by a real object moving through space in the way in which we usually think of objects moving through space. For although this idea of a continuously moving object is good enough for an approximate theory, we would discover that it would break down in a very exact theory.[4]

The natural assumption that objects, like "particles," are real things that run their course in space and time according to causal laws regardless of whether we are around to observe them or not is repudiated by quantum mechanics. This is especially significant because quantum mechanics is *the theory* of physics. It has explained successfully everything from subatomic particles to stellar phenomena. There never has been a more successful theory. It has no competition.

Therefore, when we look at the tracks in a bubble chamber, we are left with the question, "What made them?" The best answer that physicists have so far is that "particles" are actually interactions between fields. A field, like a wave, is spread out over a much larger area than a particle (a particle is restricted to one point). A field, moreover, completely fills a given space, like the gravitational field of the earth fills all of the space immediately around it.

When two fields interact with each other they interact neither gradually nor at all their areas of contact. Rather, when two fields interact, they do it *instantaneously* and at one single point in space ("instantaneously and locally"). These instantaneous and local interactions make what we call particles. In fact, according to this theory, these instantaneous and local interactions *are* "particles." The continual creation and annihilation of particles at the subatomic level is the result of the continual interaction of different fields.

This theory is called quantum field theory. Some major cornerstones of the theory were laid in 1928 by the English physicist Paul Dirac. Quantum field theory has been highly successful in predicting new types of particles and in explaining existing particles in terms of field interactions. According to this theory, a separate field is associated with each type of particle. Since only three types of particles were known in 1928, only three different fields were required to explain them. The problem today, however, is that there are over one hundred known particles, which, according to quantum field theory, require over one hundred different fields. This abundance of theoretical fields is somewhat awkward, not to mention embarrassing, to physicists whose goal is to simplify nature. Therefore, most physicists have given up the idea of a separate field existing for each type of particle.

Nevertheless, quantum field theory is still an important theory not only because it works, but also because it was the first theory to merge quantum mechanics and relativity, albeit in a limited way. All physical theories, including quantum theory, must satisfy the requirement of relativity theory that the laws of physics be independent of the state of motion of the observer. Attempts to integrate the theory of relativity with quantum theory, however, have been generally unsuccessful. Nonetheless, both relativity and quantum theory are required, and routinely used, in the understanding of particle physics. Their forced relationship is best described as strained but necessary. In this regard, one of the most successful integrations of the two is quantum field theory, although it covers only a relatively small range of phenomena.*

---

*S-Matrix theory merges quantum theory and relativity, but it provides limited information on the details of subatomic phenomena and it currently is restricted to hadron interactions (S-Matrix theory is discussed in the next chapter).

Quantum field theory is an *ad hoc* theory. That means that, like Bohr's famous specific-orbits-only model of the atom, quantum field theory is a practical but conceptually inconsistent scheme. Some parts of it don't fit together mathematically. It is a working model designed around the available data to give physicists a place to stand in the exploration of subatomic phenomena. The reason that it has been around so long is that it works so well. (Some physicists think that it may work *too* well. They fear that the pragmatic success of quantum field theory impedes the development of a consistent theory.)

Even with these well-known shortcomings, the fact is that quantum field theory is a successful physical theory, and it is premised on the assumption that *physical reality is essentially nonsubstantial.* According to quantum field theory, fields alone are real. *They* are the substance of the universe and not "matter." Matter (particles) is simply the momentary manifestations of interacting fields which, intangible and insubstantial as they are, are the only real things in the universe. Their interactions seem particle-like because fields interact very abruptly and in very minute regions of space.

"Quantum field theory" is, of course, an outrageous contradiction in terms. A quantum is an indivisible whole. It is a small piece of something, while a field is a whole area of something. A "quantum field" is the juxtaposition of two irreconcilable concepts. In other words, it is a paradox. It defies our categorical imperative that something be either *this* or *that,* but not both.

The major contribution of quantum mechanics to western thought, and there are many, may be its impact on the artificial categories by which we structure our perceptions, since ossified structures of perception are the prisons in which we unknowingly become prisoners. Quantum theory boldly states that something can be this *and* that (a wave *and* a particle).* It makes no sense to ask which of these

---

*The language of quantum theory is precise but tricky. Quantum theory does not state that something—like light, for example—can be wave-like and particle-like *at the same time.* According to Bohr's complementarity (page 103), light reveals either a particle-like aspect or a wave-like aspect depending upon the context, i.e., the experiment. It is not possible to observe both the wave-like aspect

is really the true description. Both of them are required for a complete understanding.

In 1922, Werner Heisenberg, as a student, asked his professor and friend-to-be, Niels Bohr, "If the inner structure of the atom is as closed to descriptive accounts as you say, if we really lack a language for dealing with it, how can we ever hope to understand atoms?"

Bohr hesitated for a moment and then said, "I think we may yet be able to do so. But in the process we may have to learn what the word 'understanding' really means."[5]

In human terms, it means that the same person can be good *and* evil, bold *and* timid, a lion *and* a lamb.

All of the above notwithstanding, particle physicists of necessity analyze subatomic particles as if they *were* like little baseballs that fly through space and collide with each other. When a particle physicist studies a track on a bubble-chamber photograph of a particle interaction, he assumes that it *was* made by a little moving object, and that the other tracks on the photograph likewise were made by small moving objects. In fact, particle interactions are analyzed in much the same terms that can be applied to the collision of billiard balls. Some particles collide (and are annihilated in the process) and other newly created particles come flying out of the collision area. In short, particle interactions are analyzed essentially in terms of masses, velocities, and momenta. These are the concepts of Newtonian physics and they also apply to automobiles and streetcars.

Physicists do this because they have to use these concepts if they are to communicate at all. What is available to them is usually a black photograph with white lines on it. They know that (1) according to quantum theory, subatomic particles have no independent existence of their own, (2) subatomic particles have wave-like characteristics as

---

and the particle-like aspect in the same situation. However, *both of these mutually exclusive (complementary) aspects are needed to understand "light."* In this sense, light is both particle-like and wave-like.

well as particle-like characteristics, and (3) subatomic particles actually may be manifestations of interacting fields. Nonetheless, these white lines (more patterns) lend themselves to analysis in classical terms, and so that is how particle physicists analyze them.

This dilemma, the dilemma of having to talk in classical terms about phenomena which cannot be described in classical concepts is the basic paradox of quantum mechanics. It pervades every part of it. It is like trying to explain an LSD experience. We try to use familiar concepts as points of departure, but beyond that, the familiar concepts do not fit the phenomena. The alternative is to say nothing at all.

"Physicists who deal with the quantum theory," wrote Heisenberg,

> are also compelled to use a language taken from ordinary life. We act as if there really were such a thing as an electric current [or a particle] because, if we forbade all physicists to speak of electric current [or particles] they could no longer express their thoughts.[6]

Therefore, physicists talk about subatomic particles as if they were real little objects that leave tracks in bubble chambers and have an independent ("objective") existence. This convention has been extremely productive. Over the last forty years almost one hundred particles have been discovered. They constitute what Kenneth Ford calls the particle zoo.*

The first thing to know about the particle zoo is that every particle of the same species looks exactly alike. Every electron looks exactly like every other electron. If you've seen one, you've seen them all. Likewise, every proton looks exactly like every other proton, every neutron looks exactly like every other neutron, and so on. Subatomic particles of the same type are absolutely indistinguishable.

---

*One of the finest popular books on particle physics is *The World of Elementary Particles,* by Kenneth Ford, New York, Blaisdell, 1965.

Subatomic particles of different types, however, can be recognized by their distinguishing characteristics (properties). The first distinguishing characteristic of a subatomic particle is its mass. A proton, for example, has about 1800 times more mass than an electron. (This does not necessarily mean that a proton is 1800 times larger than an electron since mass and size are not the same thing—a pound of lead and a pound of feathers have the same mass).

When physicists refer to the mass of a particle, unless they indicate otherwise, they are referring to the mass of the particle when it is at rest. The mass of a particle at rest is called its rest mass. Any mass other than a rest mass is called a relativistic mass. Since the mass of a particle increases with velocity, a particle can have any number of relativistic masses. The size of a particle's relativistic mass depends upon its velocity. For example, at 99 percent of the speed of light a particle's mass is seven times larger than it is when the particle is at rest.

At velocities above 99 percent of the speed of light particle masses increase dramatically. When the former electron accelerator at Cambridge, Massachusetts, was in operation, it received electrons from a small feeder accelerator. The electrons from the feeder accelerator were fed into the main accelerator at .99986 the velocity of light. The main accelerator then increased the velocity of these electrons to .999999996 the speed of light. This increase in velocity may look significant, but actually it is negligible. The difference between the initial velocity of the accelerated electrons and the final velocity of the accelerated electrons is the same as the difference in velocity between one automobile that can make a given trip in two hours and a faster automobile that can make the same trip in one hour fifty-nine minutes and fifty-nine seconds.[7]

The *mass* of each electron, however, increased from 60 times to as much as 11,800 times the electron rest mass! In other words, particle accelerators are misnamed. They do not increase the velocities of subatomic particles (the definition of "acceleration") as much as they increase their mass. Particle accelerators are actually particle enlargers (massifiers?).

The masses of particles, whether at rest or in motion, are measured in electron volts. An electron volt has nothing to do with electrons. An electron volt is a unit of *energy*. (It is the energy acquired by any particle with one unit of charge falling through a potential difference of one volt.) The point is that to measure something in terms of electron volts is to measure its *energy*, yet this is precisely the unit of measurement that particle physicists use to measure a particle's *mass*. For example, the rest mass of an electron is .51 million electron volts (Mev) and the rest mass of a proton is 938.2 million electron volts. The transformation of mass into energy and energy into mass is such a routine phenomenon in particle physics that particle physicists employ units of energy to designate a particle's mass.

Mass is only one particular form of energy (page 173), the energy of being. If a particle is moving it not only has energy of being (its mass) but it also has energy of motion (kinetic energy). Both types of energy can be used to create new particles in a particle collision.*

Often it is easier to compare a particle's mass with the lightest massive particle, the electron, instead of referring to the number of electron volts it contains. This arrangement makes the mass of an electron one and the mass of a proton, for example, 1836.12. Using this system, the mass of any particle tells immediately how much heavier it is than an electron. This is the system that is used in the table at the back of the book.

When physicists listed all the known particles by the order of their masses, from the lightest to the heaviest, they discovered that subatomic particles fall into roughly three categories: the light-weight particles, the medium-weight particles, and the heavy-weight particles. When it came to naming these categories, however, they unaccountably lapsed into Greek again. The group of light-weight particles they called "leptons," which is Greek for "the light ones." The group

---

*Einstein's formula $E = mc^2$ says that mass is energy: energy is mass. Therefore, strictly speaking, mass is not a particular form of energy. *Every form of energy is mass.* Kinetic energy, for example, is mass. As we speed up a particle, and hence give it energy, ΔE, it gains mass, Δm, in exactly the amount required: $\Delta E = (\Delta m)c^2$. Wherever energy goes, mass goes.

of medium-weight particles they called "mesons" (maze'ons), which is Greek for "the medium-sized ones." The group of heavy-weight particles they called "baryons" (bary'ons), which is Greek for "the heavy ones." Why physicists did not call these new groups "light," "medium," and "heavy" is one of the unanswerable questions of physics.*

Since the electron is the lightest material particle, it is, of course, a lepton. The proton is a heavy-weight particle (a baryon), although it is the lightest of the heavy-weight particles. Most subatomic particles are classified in this way, but not all of them, which brings us to a phenomenon of particle physics which, like much of quantum mechanics, escapes the bounds of concept. A few particles do not fit into the lepton-meson-baryon framework. Some of them are well known (like the photon) and others have been theorized but not discovered yet (like the graviton). All of them have in common the fact that they are *massless particles.*

"Wait a minute," we exclaim. "What is a massless particle?"

"A massless particle," says Jim de Wit, who has studied this phenomenon, "is a particle that has zero rest mass. All of its energy is energy of motion. When a photon is created, it instantly is traveling at the speed of light. It cannot be slowed down (it has no mass to slow) and it cannot be speeded up (nothing can travel faster than the speed of light)."

"Massless particle" is an awkward translation from mathematics to English. Physicists know exactly what they mean by a massless particle. A "massless particle" is the name they give to an element in a mathematical structure. What that element represents in the real

---

*Physicists no longer use the terms leptons, mesons, and baryons to refer to particle mass alone. These terms now refer to classes of particles which are defined by several properties in addition to mass. For example, the tau particle (τ), which was discovered by a joint team from the Stanford Linear Accelerator Center (SLAC) and the Lawrence Berkeley Laboratory (LBL) in 1975, seems to be a lepton even though it has more mass than the heaviest baryon! Similarly, the D particles, also discovered by a joint SLAC/LBL team (in 1976) are mesons even though *they* have more mass than the tau particle.

world, however, is not so easy to describe. In fact, it is impossible because the definition of an object (like a "particle") is something that has mass.

Zen Buddhists have developed a technique called the *koan* which, along with meditation, produces changes in our perceptions and understanding. A *koan* is a puzzle which cannot be answered in ordinary ways because it is paradoxical. "What is the sound of one hand clapping?" is a Zen *koan*. Zen students are told to think unceasingly about a particular *koan* until they know the answer. There is no single correct answer to a *koan*. It depends upon the psychological state of the student.

Paradoxes are common in Buddhist literature. Paradoxes are the places where our rational mind bumps into its own limitations. According to eastern philosophy in general, opposites, such as good-bad, beautiful-ugly, birth-death, and so on, are "false distinctions." One cannot exist without the other. They are mental structures which we have created. These self-made and self-maintained illusions are the sole cause of paradoxes. To escape the bonds of conceptual limitation is to hear the sound of one hand clapping.

Physics is replete with *koans,* i.e., "picture a massless particle." Is it a coincidence that Buddhists exploring "internal" reality a millennium ago and physicists exploring "external" reality a millennium later both discovered that "understanding" involves passing the barrier of paradox?

The second characteristic of a subatomic particle is its charge. Every subatomic particle has a positive, a negative, or a neutral charge. Its charge determines how the particle will behave in the presence of other particles. If a particle has a neutral charge, it is utterly indifferent to other particles, regardless of what charge they may have. Particles with positive and negative charges, however, behave quite differently toward each other. Positively and negatively charged particles are attracted to particles with the opposite sign and repelled by particles with the same sign. Two positively charged particles, for example, find each other's company quite repulsive and immediately put as much distance between themselves as possible. The same is true of two negatively charged particles. A negatively charged particle and a positively charged particle, on the other hand, are irresistibly attracted to each other, and they immediately move toward one another if they are able to do so.

This dance of attraction and repulsion between charged particles is called the electromagnetic force. It enables atoms to join together to form molecules and it keeps negatively charged electrons in orbit around positively charged nuclei. At the atomic and molecular level it is the fundamental glue of the universe.

Electric charge comes only in one fixed amount. A subatomic particle can have no electrical charge (neutral), or one unit of electrical charge (either positive or negative), or, in certain instances, two units of electrical charge, but nothing in between. There is no such thing as a particle with one and one fourth units of electrical charge, or a particle with 1.7 units of electrical charge. Every subatomic particle has either one whole unit of electrical charge, two whole units of electrical charge, or no electrical charge at all. In other words, like energy (Planck's discovery) electrical charge is "quantized." It comes in chunks. In the case of electrical charge, all of the chunks are the same size. Why this is so is one of THE unanswered questions in physics.*

When the characteristic of charge is added to the characteristic of mass, a particle personality, so to speak, begins to emerge. An "electron," for example, is the only subatomic particle with a rest mass of .51 million electron volts and a negative charge. With this information a particle physicist knows not only how massive an electron is, he also knows how it will interact with other particles.

The third characteristic of a subatomic particle is its spin. Subatomic particles spin about a theoretical kind of axis like a spinning top. One

---

*This peculiar aspect of electrical charge appears to be connected to the unknown properties of quarks and/or magnetic monopoles.

big difference between a spinning top and a spinning particle, however, is that a top can spin either faster or slower, but a subatomic particle *always* spins at exactly the same rate. Every electron, for example, always spins at exactly the same rate as every other electron.

The rate of spin is such a fundamental characteristic of a subatomic particle that if it is altered, the particle itself is destroyed. That is, if the spin of a particle is altered, the particle in question is changed so fundamentally that it no longer can be considered an electron, or a proton, or whatever it was before we altered its spin. This makes us wonder whether all of the different "particles" might be just different states of motion of some underlying structure or substance. This is the basic question of particle physics.

Every phenomenon in quantum mechanics has a quantum aspect which makes it "discontinuous." This is also true of spin. Spin is quantized just like energy and charge. It comes in chunks. Like charge, all of the chunks are the same size. In other words, when a spinning top slows down, its rotation does not diminish smoothly and continuously, but in a series of tiny steps. These steps are so small and close together that it is impossible to observe them. The top appears to spin more and more slowly until it stops spinning altogether, but actually, the process is very jerky.

It is as if the top could spin, by some strange law that nobody understands, only at 100 revolutions per minute, 90 revolutions per minute, 80 revolutions per minute, and so on, with absolutely no exceptions in between. If our hypothetical top wants to spin slower than 100 revolutions per minute, it must jump all the way down to the next slower speed of 90 revolutions per minute. This is analogous to the situation with subatomic particles except that (1) particular types of particles forever spin at the same speed, and (2) the spin of subatomic articles is calculated in terms of angular momentum.

Angular momentum depends upon the mass, size, and rate of rotation of a spinning object. More of any one of these properties increases the angular momentum of the object. In general, angular momentum is the strength of the rotation or, put another way, the effort required to stop the rotation. The more angular momentum an object has, the more effort is required to stop it from spinning. A spinning top does not have much angular momentum, because it is small and it has little mass. A merry-go-round, in comparison, has an enormous angular momentum, not because it rotates very fast, but because it is large and it has so much mass.

Now that you understand spin, forget everything that you have just learned except the bottom line (angular momentum). Every sub-atomic particle has a fixed, definite, and known angular momentum, but *nothing is spinning!* If you don't understand, don't worry. Physicists don't understand these words, either. They just use them. (If you try to *understand* them, they become a *koan*).*

The angular momentum of a subatomic particle is fixed, definite, and known. "But," wrote Max Born,

> one should not imagine that there is anything in the nature of matter actually rotating.[8]

Said another way, the "spin" of a subatomic particle involves "The idea of a spin without the existence of something spinning . . ."[9] Even Born had to admit that this concept is "rather abstruse."[10] (Rather!?) Nonetheless, physicists use this concept because subatomic particles *do* behave as if they have angular momentum and that angular momentum has been determined to be fixed and definite in each case. Because of this, in fact, "spin" is one of the major characteristics of subatomic particles.

*The quantitative (mathematical) description of particle spin is not any more *understandable* than the nonquantitative description. Dr. Felix Smith, Head of Molecular Physics, Stanford Research Institute, once related to me the true story of a physicist friend who worked at Los Alamos after World War II. Seeking help on a difficult problem, he went to the great Hungarian mathematician, John von Neumann, who was at Los Alamos as a consultant.

"Simple," said von Neumann. "This can be solved by using the method of characteristics."

After the explanation, the physicist said, "I'm afraid I don't understand the method of characteristics."

"Young man," said von Neumann, "in mathematics you don't understand things, you just get used to them."

The angular momentum of a subatomic particle is based upon our old friend, Planck's constant (page 56). Remember that Planck's constant, which physicists call "the quantum of action," was the discovery that set into motion the revolution of quantum mechanics. Planck discovered that energy is emitted and absorbed not continuously, but in small packages which he called quanta. Since that initial discovery, Planck's constant, which represents the quantized nature of energy emission and absorption, has appeared again and again as an essential element in the understanding of subatomic phenomena. Five years after Planck's discovery, Einstein used Planck's constant to explain the photoelectric effect, and later still he used it to determine the specific heat of solids, an area far removed from Planck's original study of black-body radiation. Bohr discovered that the angular momentum of electrons as they orbit atomic nuclei is a function of Planck's constant, de Broglie used Planck's constant to calculate the wavelength of matter waves, and it is a central element in Heisenberg's uncertainty principle.

As important as Planck's constant is in the realm of subatomic particles, however, it is entirely unobservable in the world at large. This is because the size of the packages by which energy is emitted and absorbed is so small that energy at our gross level appears to be one continuous flow. Similarly, because the indivisible unit of angular momentum is so small, it, too, cannot be observed in the macroscopic world. A spectator swiveling in his chair at a tennis match has 1000000000000000000000000000000000 (10^{33}) times more angular momentum than an electron. Put another way, a change of one penny in the gross national product of the United States is a disturbance more than a *billion billion* times greater than a change by one unit of the spectator's angular momentum.[11]

Instead of writing out the actual angular momentum of a subatomic particle, physicists usually indicate the spin of a subatomic particle by comparing it to the spin of a photon, whose spin they call one. This system has revealed yet another unexplainable pattern of subatomic phenomena. Entire families of particles have similar spin characteristics. The entire family of leptons, the light-weight particles, for example, has a spin of $1/2$, which means that they *all* have an angular momentum which is $1/2$ of a photon's angular momentum. The same is true for the entire family of baryons, the heavy-weight particles. The mesons also have peculiar spin characteristics. They spin in such a way that their angular momenta is always either 0, 1, 2, 3, etc. in relation to the angular momentum of a photon, but nothing in between (0 = no spin, 1 = the same angular momentum as a photon, 2 = twice the angular momentum of a photon, etc.). The spin characteristics of all of the families and all of the particles are in the table at the back of the book.

The values of a particle's charge, spin and other major characteristics are represented by specific numbers. These numbers are called quantum numbers. Every particle has a set of quantum numbers which identify it as a particular type of particle.* Every particle of a particular type has the same set of quantum numbers as every other particle of the same type. Every electron, for example, has the same quantum numbers as every other electron. An electron's quantum numbers, however, distinguish it from protons, all of which also have the same quantum numbers. Individual particles don't have much personality. In fact, they don't have any personality at all.

When Dirac imposed the requirements of relativity on quantum theory, his formalism indicated the existence of a particular positively charged particle. Since the only positively charged particle known in those days (1928) was the proton, Dirac, and most other physicists, assumed that his theory had accounted (mathematically) for the proton. (His theory even was criticized for yielding the "wrong" proton mass.)

Upon closer examination, however, it became evident that Dirac's theory described not the proton, but an entirely different particle. Dirac's new particle was like an electron except that its charge and some of its other major properties were exactly opposite to those of an electron.

*The basic quantum numbers are spin, isotopic spin, charge, strangeness, charm, baryon number, and lepton number.

In 1932, Carl Anderson, at Cal Tech (who hadn't heard of Dirac's theory) actually discovered this new particle and called it a positron. Physicists later discovered that every particle has a counterpart which is exactly like it but opposite in several major respects. This new class of particles was called anti-particles. An anti-particle, despite its name, is a particle. (The anti-particle of an anti-particle is another particle.)

Some particles have other particles as anti-particles (for example, a positive pi meson is the anti-particle of a negative pi meson, and the other way round). A few particles are their own anti-particles (like the photon). All of the particles and their anti-particles are in the table at the back of the book.

The meeting of a particle and its anti-particle is always spectacular. Whenever a particle and its anti-particle meet, they annihilate each other! When an electron meets a positron, for example, both of them disappear and in their place are two photons which instantly depart the scene at the speed of light. The particle and the anti-particle literally disappear in a puff of light. Conversely, particles and anti-particles can be created out of energy and always in pairs.

The universe is made of both particles and anti-particles. Our part of it, however, is made almost entirely of regular particles which combine into regular atoms to make regular molecules, which make regular matter, which is what we are made of. Physicists speculate that in other parts of the universe anti-particles combine into anti-atoms to make anti-molecules, which make anti-matter, which is what anti-people would be made of. There are no anti-people in our part of the universe because, if there were, they all long since have disappeared in a flash of light.

Leptons, mesons, baryons, mass, charge, spin, and anti-particles are some of the concepts that physicists use to categorize subatomic phenomena when they momentarily assume that subatomic particles are real objects that move through space and time. These concepts are useful, but only in a limited context. That context is when physicists, for convenience, pretend, as we all do, that dancers can exist apart from a dance.

1

🌀

The Dance

The dance of subatomic particles never ends and it is never the same. However, physicists have found a way to diagram the parts of it that interest them.

The simplest drawing of any type of movement is a space map. A space map shows the location of things in space. The map, for example (first drawing, next page), shows the positions of San Francisco, California, and Berkeley, California. The vertical axis is the north-south axis, as on any map, and the horizontal axis is the east-west line. The map also shows the path of a helicopter flying between San Francisco and Berkeley and, on a greatly enlarged basis, it shows the path of a proton traveling around the cyclotron at the Lawrence Berkeley Laboratory.

Like all road maps, this space map is two-dimensional. It shows how far north (the first dimension) and how far east (the second dimension) Berkeley is of San Francisco. It does not show the altitude of the helicopter (the third dimension) and it does not indicate how much time (the fourth dimension) the flight from San Francisco to Berkeley required. If we want to show the time involved in the San Francisco–Berkeley flight we must draw a space-time map.

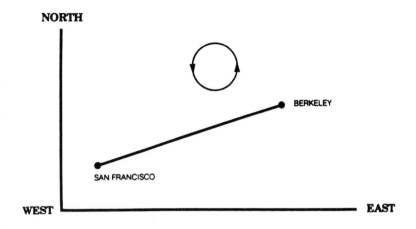

A space-time map shows the positions of things in space and it also shows their positions in time. The vertical axis on a space-time map is the time axis. Space-time maps are read from the bottom up because the passage of time is represented by movement up the time axis. The horizontal axis of a space-time map is the space axis which shows the movement of objects in space. The path traced by an object on a space-time map is called its "world line." For example, the space-time map below shows the same flight from San Francisco to Berkeley.

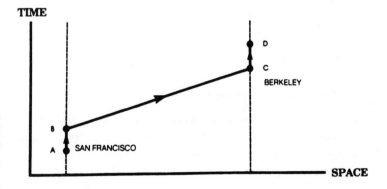

Initially the helicopter is sitting on the ground in San Francisco. Its world line is vertical, because, although it is not moving in space, it is moving in time. A to B is the world line of the helicopter while it sits on its pad in San Francisco. When the helicopter takes off for Berkeley it moves forward both in time and space, and its world line traces the path on the space-time map between B and C. When it lands in Berkeley its world line is vertical once again because it no longer moves in space, but, like all things, it continues to move in time (C to D). The arrowheads show which direction and helicopter is moving. It can move backward and forward in space but, of course, it only can move forward in time. The dashed lines show the world lines of San Francisco and Berkeley which do not move in space at all except during California earthquakes.

Physicists use similar space-time maps to diagram particle interactions. Below is a space-time diagram of an electron emitting a photon.

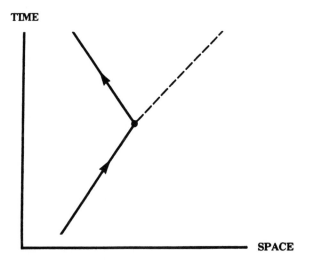

Starting at the bottom, an electron moves through space with a certain velocity. At the point in space and time indicated by the dot, it emits a photon. The photon flies off at the speed of light to the right and the electron, its momentum affected by the emission of the photon, alters course and moves off more slowly to the left.

In 1949, Richard Feynman discovered that space-time maps like these have an exact correspondence with mathematical expressions which give the probabilities of the interactions that they depict. Feynman's discovery was an extension of Dirac's 1928 theory and it helped to evolve that theory into the quantum field theory that we know today. Therefore, this type of diagram is sometimes called a Feynman diagram.*

Here is a Feynman diagram of a particle/anti-particle annihilation. An electron on the left approaches an anti-electron (a positron) which is coming from the right. At their point of contact, indicated by the dot, they mutually annihilate each other and two photons are created which depart in opposite directions at the speed of light.

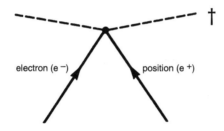

A happening in the subatomic world is called an "event." Events are indicated in Feynman diagrams by dots. *Every subatomic event is*

---

*Original diagrams of this sort were space-time diagrams. However, Feynman also discovered that momentum-energy space descriptions, which are complementary to space-time descriptions, more closely approximate the actual conditions of a collision experiment. The basic concept of space-time descriptions and momentum-energy space descriptions is the same except that momentum-energy space descriptions deal with the momenta and energies of the particles involved instead of their space-time co-ordinates. The diagrams of both space-time descriptions and momentum-energy space descriptions are similar except that diagrams depicting momentum-energy space descriptions can be rotated, as we shall see. Accurately speaking, the remaining Feynman diagrams in this book depict momentum-energy space descriptions unless they are specifically identified as space-time diagrams.

†A detailed analysis of the dot in this particular diagram would reveal a two-step process in which first one photon and then the other is emitted. Technically,

marked by the annihilation of the initial particles and the creation of new ones. This is true for every event and not only those involving particles and anti-particles.

With this in mind, we now can look at the particle diagram on page 238 again, and see it in a different light. Instead of saying that an electron moving through space emitted a photon which changed its (the electron's) momentum, we can say as well that an electron moving through space emitted a photon and went out of existence at that point! A new electron was created in this process and *it* departed the scene with a new momentum. There is no way of knowing if this interpretation is correct or not because all electrons are identical. However, it is simpler and more consistent to assume that the original particle was annihilated and a new particle was created. The indistinguishability of subatomic particles makes this possible.

On the next page is a Feynman diagram of the process that we discussed on pages 218–19.

A negative pi meson collides with a proton and the two particles are annihilated. Their energy of being (mass) and energy of motion create two new particles, a lambda particle and a neutral K meson. These two new particles are unstable and live less than a billionth of a second before they decay into other particles (actual decay times are in the table at the back of the book). The neutral K meson decays into a positive pi meson and a negative pi meson. The lambda particle, and this is the interesting part, decays into the original two particles (a negative pi meson and a proton)! It is as if we smash two toy automobiles together and instead of shattering into bits and pieces, they come apart into more toy automobiles, some of which are as large as the originals.

Subatomic particles forever partake of this unceasing dance of annihilation and creation. In fact, subatomic particles *are* this unceasing dance of annihilation and creation. This twentieth-century discovery, with all its psychedelic implications, is not a new concept. In fact,

---

diagrams with more than three lines connected to the same vertex are called Mandelsten diagrams.

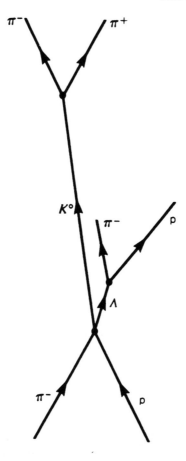

it is very similar to the way that much of the earth's population, including the Hindus and the Buddhists, view their reality.

Hindu mythology is virtually a large-scale projection into the psychological realm of microscopic scientific discoveries. Hindu deities such as Shiva and Vishnu continually dance the creation and destruction of universes while the Buddhist image of the wheel of life symbolizes the unending process of birth, death, and rebirth which is a part of the world of form, which is emptiness, which is form.

Imagine that a group of young artists have founded a new and revolutionary school of art. Their paintings are so unique that they have come to share them with the curator of an old museum. The curator regards the new paintings, nods his head, and disappears into the vaults of the museum. He returns carrying some very old paintings, which he places beside the new ones. The new art is so similar to the old art that even the young artists are taken aback. The new revolutionaries, in their own time and in their own way, have rediscovered a very old school of painting.

Let us look again at the Feynman diagram of an electron-positron annihilation. Suppose that we use the arrowhead to indicate which is the particle (the electron) and which is the anti-particle (the positron) by making the arrowheads that point up indicate the particles and the arrowheads that point down indicate the anti-particles. That would make the diagram on page 239 look like the drawing below.

Naturally, time, as we experience it, only travels in one direction, forward, and that is up on a space-time diagram. Nonetheless, this simple convention would give us an easy way of telling particles from anti-particles. World lines that appear to move forward in time would belong to particles and world lines that appear to move backward in time would belong to anti-particles. (Photons would not have arrowheads because they are their own anti-particles.)

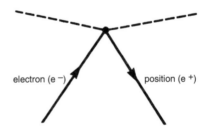

electron (e⁻)   position (e⁺)

Feynman demonstrated in 1949 that this convention is more than an artistic device. He discovered that a positron field propagating forward in time is mathematically the same as an electron field propagating backward in time! In other words, according to quantum field theory, *an anti-particle is a particle moving backward in time.* An anti-particle does not *have* to be considered as a particle moving backward in time, but that is the simplest and most symmetric way of viewing anti-particles.

For example, because the arrowheads distinguish the particles from the anti-particles, we can twist the original Feynman diagram around into any position that we choose and still be able to distinguish the one from the other. Here are some different ways that we can twist the original Feynman diagram.

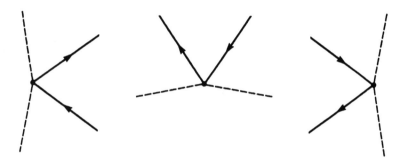

Each of these variations is a separate diagram and represents a particle/anti-particle interaction.* By twisting the original diagram completely around we can represent every possible interaction between an electron, a positron, and two photons. The precision, simplicity, and symmetry of Feynman diagrams make them a special type of poetry.

On the next page is a space-time diagram of *two events.* A collision between two photons (at B) creates an electron-positron pair and, subse-

---

*These three interactions are: *left,* a photon and an electron annihilate and create a photon and an electron (electron-photon scattering); *middle,* two photons annihilate to create a positron and an electron (positron-electron pair creation); and, *right,* a positron and a photon annihilate to create a positron and a photon (positron-photon scattering).

quently, an electron and a positron annihilate each other and create two photons (at A). (The left half of this diagram, the interaction at A, is the same as the electron-positron annihilation on page 242).

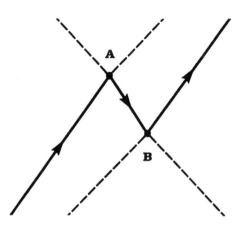

Ordinarily we would interpret these events as follows: Two photons collide in the lower right of the diagram producing an electron-positron pair. The electron flies off to the right while the positron flies off to the left where it meets another electron which has entered the diagram from the lower left. There they mutually annihilate and create two photons which depart in opposite directions.

The preferred interpretation of quantum field theory, however, is much simpler. In it there is only *one* particle. That particle, an electron, enters the diagram from the lower left and travels forward in time and space until it emits two photons at A. This causes it to reverse its direction in time. Traveling backward in time as a positron it absorbs two photons at B, reverses its direction in time again, and again becomes an electron. Instead of three particles there is only one particle which, moving from left to right, travels first forward in time, then backward in time, and then forward in time again.

This is the static type of space-time picture described in Einstein's theory of relativity (page 168). If we could survey an entire span of time as we can survey an entire region of space, we would see that events do not unfold with the flow of time but present themselves complete, like a finished painting on the fabric of space-time. In such a picture movements backward and forward in time are no more significant than movements backward and forward in space.

The illusion of events "developing" in time is due to our particular type of awareness which allows us to see only narrow strips of the total space-time picture one at a time. For example, suppose that we place a piece of cardboard with a narrow strip cut out of it over the diagram on page 244 so that all we can see of the interaction is what is visible through the cut-out. If we move the cardboard slowly upward, starting at the bottom, our restricted view discovers a series of events, each one happening after the other.

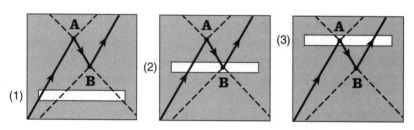

First, we see three particles, two photons entering our view on the right and an electron entering from the left (1). Next, we see the photons collide to produce an electron-positron pair, the electron flying off to the right and the positron flying off to the left (2). Finally, we see the newly created positron meet the original electron to create two new photons (3). Only when we remove the entire cardboard (which was an artificial construction anyway) can we see the complete picture.

"In space-time," wrote de Broglie,

> everything which for each of us constitutes the past, the present, and the future is given in block. . . . Each observer, as his time passes, discovers, so to speak, new slices of space-time which appear to him as successive aspects of the material world, though in reality the ensemble of events constituting space-time exist prior to his knowledge of them.[1]

"Wait a minute," says Jim de Wit to a passing particle physicist. "It is easy to talk of movement backward and forward in time, but I never have experienced going backward in time. If particles can travel backward in time, why can't *I* travel backward in time?"

The answer which physicists gave to this question is actually quite simple: There is a growing tendency in any closed part of the universe, their explanation goes, for disorder (called "entropy") to expand at the price of order (called "negentropy"). Suppose, for example, that we deposit a drop of black ink into a glass of clear water. Initially its presence is quite ordered. That is, all of the molecules of ink are located in one small area and are clearly segregated from the molecules of clear water.

As time passes, however, natural molecular motion will cause the black ink molecules steadily to intersperse with the clear water molecules until they are distributed evenly throughout the glass, resulting in a murky homogeneous liquid with no structure or order whatever—only a bland uniformity (maximal entropy).

Experience has taught us to associate increasing entropy with the forward movement of time. If we see a movie of a glass of murky water becoming clearer and clearer until all of the foreign substance in it collects into one small drop at the top, we know at once that the film is running backward. Of course, it is theoretically possible for this to happen, but it is so improbable that it simply never (probably) will happen. In short, time "flows" in the direction of high probability, which is the direction of increasing entropy.

The theory of growing disorder, or "increasing entropy," is called the second law of thermodynamics. The second law of thermodynamics is statistical. That means it won't work unless there are many entities in a given situation to apply it to. Generally speaking, individual subatomic particles are conceived as such conceptually isolated, short-lived entities that the second law of thermodynamics does not apply to them.*,† It does apply, however, to molecules, which are

---

*The Hagedorn theory of very-high-energy collisions uses the second law of thermodynamics.

†Time reversability exists *in potentia,* i.e., while the particles are represented by propagating wave functions. Time irreversability is an artifact of the measurement process.

quite complex compared to subatomic particles; to living cells, which are more complex than molecules; and to people, who are made of billions of cells. It is only at the subatomic, or quantum, level that the forward flow of time loses its significance.

However, there is speculation, and some evidence, that consciousness, at the most fundamental levels, *is* a quantum process. The dark-adapted eye, for example, can detect a single photon. If this is so then it is conceivable that by expanding our awareness to include functions which normally lie beyond its parameters (the way yogis control their body temperature and pulse rate) we can become aware of (experience) these processes themselves. *If,* at the quantum level, the flow of time has no meaning, and *if* consciousness is fundamentally a similar process, and *if* we can become aware of these processes within ourselves, then it also is conceivable that we can experience timelessness.

If we can experience the most fundamental functions of our psyche, and if they are quantum in nature, then it is possible that the ordinary conceptions of space and time might not apply to them at all (as they don't seem to apply in dreams). Such an experience would be difficult to describe rationally ("Infinity in a grain of sand/And eternity in an hour"), but it would be very real, indeed. For this reason, reports of time distortion and timelessness from gurus in the East and psychotropic drug users in the West ought not, perhaps, to be discarded preemptorily.

Subatomic particles do not just sit around being subatomic particles. They are beehives of activity. An electron, for example, constantly is emitting and absorbing photons. These photons are not full-fledged photons, however. They are a now-you-see-it-now-you-don't variety. They are exactly like real photons except that they don't fly off on their own. They are re-absorbed by the electron almost as soon as they are emitted. Therefore, they are called "virtual" photons. ("Virtual" means "being so in effect or essence, although not in actual fact.") They are virtually photons. The only thing that keeps them from being full-fledged photons is their abrupt re-absorption by the electron that emits them.*

In other words, first there is an electron, then there is an electron and a photon, and then there is an electron again. This situation is, of course, a violation of the conservation law of mass-energy. The conservation law of mass-energy says, in effect, that we cannot get something for nothing. According to quantum field theory, however, we do get something for nothing, but only for about one thousand trillionth (10^{-15}) of a second.† The reason that this can happen, according to the theory, is the famous Heisenberg uncertainty principle.

The Heisenberg uncertainty principle, as it originally was formulated, says that the more certain we are of the position of a particle, the less certain we can be about its momentum, and the other way round. We can determine its position precisely, but in that case we cannot determine its momentum at all. If we choose to measure its momentum precisely, then we will not be able to know where it is located (page 122).

In addition to the reciprocal uncertainty of position and momentum, there also is a reciprocal uncertainty of time and energy. The less uncertainty there is about the time involved in a subatomic event, the more uncertainty there is about the energy involved in the event (and the other way round). A measurement as accurate as one thousand trillionth of a second leaves very little uncertainty about the time involved in the emission and absorption of a virtual photon. It does, however, cause a specific uncertainty about how much energy was involved. Because of this uncertainty, the balance books kept by the

---

*From one point of view virtual photons differ from real ones in that their rest mass is not zero: only zero rest mass photons can escape. There are two ways of looking at virtual photons mathematically. In the first (old-fashioned perturbation theory), the mass of a virtual particle is the same as the mass of a real particle, but energy is not conserved. In the second (Feynman perturbation theory) energy-momentum is exactly conserved, but the virtual particles do not have physical mass.

†In a typical atomic process; high-energy virtual photons have even shorter lifetimes.

conservation law of mass-energy are not upset. Said another way, the event happens and is over with so quickly that the electron can get away with it.

It is as if the policeman who enforces the conservation law of mass-energy turns his back on violations if they happen quickly enough. However, the more flagrant the violation, the more quickly it must happen.

If we provide the necessary energy for a virtual photon to become a real photon without violating the conservation law of mass-energy, it does just that. That is why an excited electron emits a real photon. An excited electron is an electron that is in an energy level higher than its ground state. An electron's ground state is its lowest energy level where it is as close to the nucleus of an atom as it can get. The only photons that electrons emit when they are in their ground state are virtual photons which they immediately re-absorb so as not to violate the conservation law of mass-energy.

An electron considers the ground state to be its home. It doesn't like to leave home. In fact, the only time it leaves its ground state is when it literally is pushed out of it with extra energy. In that case, the electron's first concern is to get back to its ground state (provided that it hasn't been pushed so far from the nucleus that, in effect, it becomes a free electron). Since the ground state is a low-energy state, the electron must lose its excess energy before it can return to it. Therefore, when an electron is at an energy level higher than its ground state, it jettisons its excess energy in the form of a photon. The jettisoned photon is one of the electron's virtual photons that suddenly finds itself with enough energy to keep going without violating the conservation law of mass-energy, and it does. In other words, one of the electron's virtual photons suddenly is "promoted" to a real photon. The amount of energy (the frequency) of the promoted photon depends upon how much excess energy the electron had to jettison. (The discovery that electrons emit only photons of certain energies and no others is what made the quantum theory a *quantum* theory). Electrons are always surrounded by a swarm of virtual photons.*

If two electrons come close enough to each other, close enough so that their virtual-photon clouds overlap, it is possible that a virtual photon that is emitted from one electron will be absorbed by the other electron. Below is a Feynman diagram of a virtual photon being emitted by one electron and absorbed by another electron.

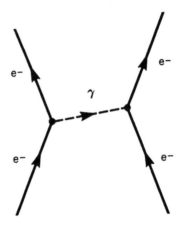

The closer the electrons come to each other, the more this phenomenon occurs. Of course, the process is two-way with both electrons absorbing virtual photons that were emitted by the other.

This is how electrons repel each other. The closer two electrons come, the more virtual photons they exchange. The more virtual photons they exchange, the more sharply their paths are deflected. The "repulsive force" between them is simply the cumulative effect of these exchanges of virtual photons, the number of which increases at close range and decreases at a distance. According to this theory, there is no such thing as action-at-a-distance—only more and fewer exchanges of virtual photons. These interactions (absorptions and

*There are other virtual particles in the cloud of virtual particles surrounding an electron, but photons are the most common among them.

emissions) happen on location, so to speak, right there where the particles involved are located.*

The mutual repulsion of two particles of the same charge, like two electrons, is an example of an electromagnetic force. In fact, according to quantum field theory, an electromagnetic force *is* the mutual exchange of virtual photons. (Physicists like to say that the electromagnetic force is "mediated" by photons.) Every electrically charged particle continually emits and re-absorbs virtual photons and/or exchanges them with other charged particles. When two electrons (two negative charges) exchange virtual photons, they repulse each other. The same thing happens when two protons (two positive charges) exchange virtual photons. When a proton and an electron (a positive charge and a negative charge) exchange virtual photons they attract each other.

Therefore, since the development of quantum field theory, physicists generally have substituted the word "interaction" for the word "force." (An interaction is when anything influences anything else). In this context—a mutual exchange of virtual photons—it is a more precise term than "force," which labels that which happens between electrons but does not say anything about it. That part of quantum field theory (Dirac's original part) which deals with electrons, photons, and positrons is called quantum electrodynamics.

Virtual photons, even if they were charged particles, would not be visible in a bubble chamber because of their extremely short lives. Their existence is inferred mathematically. Therefore, this extraordinary theory, that particles exert a force on each other by exchanging other particles, clearly is a "free creation" of the human mind (page 9). It is not necessarily how nature "really is," it is only a mental construc-

*However, the essence of quantum mechanics seems to demand a non-dynamic "action-at-a-distance" operating faster than light. A good example of this is the Pauli exclusion principle, which indicates a correlation between the motions

tion which correctly predicts what nature probably is going to do next. There might be, and probably are, other mental constructs that can do as good a job as this one, or better (although physicists have not been able to think of them). The most that we can say about this or any other theory is not whether it is "true" or not, but only whether it works or not; that is, whether it does what it is supposed to do.

Quantum theory is supposed to predict the probabilities of given subatomic phenomena to occur under certain circumstances. Even though quantum field theory as a whole is not totally consistent, the pragmatic reality is that it *works*. There is a Feynman diagram for every interaction, and every Feynman diagram corresponds to a mathematical formula which precisely predicts the probability of the diagrammed interaction to happen.*

In 1935, Hideki Yukawa, a graduate student in physics, decided to apply the new virtual particle theory to the strong force.

The strong force is the force that keeps atomic nuclei together. It has to be strong because the protons, which along with the neutrons make up the nucleus of an atom, naturally repel each other. Being particles of like sign (positive), protons want to be as far away from each other as they can get. This is because of the electromagnetic force between them. However, within the nucleus of an atom, these mutually repulsive protons not only are kept in close proximity, but they also are bound together very tightly. Whatever is binding these protons together into a nucleus, physicists reasoned, must be a very strong force compared to the electromagnetic force, which works against it. Therefore, they decided to call the strong force, naturally, the "strong force."

The strong force is well named because it is one hundred times stronger than the electromagnetic force. It is the strongest force

---

of two electrons over and above the exchange of virtual "signal" photons. (Other examples—EPR and Bell's theorem—are discussed in a later chapter.)

*In fact, there usually is an infinite series of Feynman diagrams for every interaction.

known in nature. Like the electromagnetic force, it is a fundamental glue. The electromagnetic force holds atoms together externally (with each other to form molecules) and internally (it binds electrons to their orbits around atomic nuclei). The strong force holds the nucleus itself together.

The strong force is somewhat musclebound, so to speak. Although it is the strongest force known in nature, it also has the shortest range of all the forces known in nature. For example, as a proton approaches the nucleus of an atom it begins to experience the repulsive electromagnetic force between itself and the protons within the nucleus. The closer the free proton gets to the protons in the nucleus, the stronger the repulsive electromagnetic force between them becomes. (At one third the original distance, for example, the force is nine times as strong). This force causes a deflection in the path of the free proton. The deflection is a gentle one if the proton is distant from the nucleus and very pronounced if the proton should come close to the nucleus.

However, if we push the free proton to within about one ten-trillionth (10^{-13}) of a centimeter of the nucleus, it suddenly is sucked *into* the nucleus with a force one hundred times more powerful than the repulsive electromagnetic force. One ten-trillionth of a centimeter is about the size of the proton itself. In other words, the proton is relatively unaffected by the strong force, even at a distance only slightly greater than its own magnitude. Closer than that, however, and it is completely overpowered by the strong force.

Yukawa decided to explain this powerful but very short-range "strong" force in terms of virtual particles.

The strong force, theorized Yukawa, is "mediated" by virtual particles like the electromagnetic force is "mediated" by virtual photons. According to Yukawa's theory, just as the electromagnetic force *is* the exchange of virtual photons, the strong force *is* the exchange of another type of virtual particle. Just as electrons never sit idle, but constantly emit and re-absorb virtual photons, so nucleons are not inert, but constantly emit and re-absorb their own type of virtual particles.

A "nucleon" is a proton or a neutron. Both of these particles are called nucleons, since both of them are found in the nuclei of atoms. They are so similar to each other that a proton, roughly speaking, can be considered as a neutron with a positive charge.

Yukawa knew the range of the strong force from the results of published experiments. Assuming that the limited range of the strong force was identical to the limited range of a virtual particle emitted from a nucleon in the nucleus, he calculated how much time such a virtual particle would require, at close to the speed of light, to go that distance and return to the nucleon. This time calculation allowed him to use the uncertainty relation between time and energy to calculate the energy (mass) of his hypothetical particle.

Twelve years and one case of mistaken identity later, physicists discovered Yukawa's hypothetical particle.* They called it a meson. *An entire family of mesons*, it later was discovered, are the particles which nucleons exchange to constitute the strong force. The particular meson which physicists discovered first, they called a pion. "Pion" is short for pi (pronounced "pie") meson. Pions come in three varieties: positive, negative, and neutral.

In other words, a proton, like an electron, is a beehive of activity. Not only does it continually emit and re-absorb virtual photons, which makes it susceptible to the electromagnetic force, it also emits and re-absorbs virtual pions, which makes it susceptible to the strong force as well. (Particles which do not emit virtual mesons, like electrons, for example, are not affected at all by the strong force).

When an electron emits a virtual photon which is absorbed by another particle, the electron is said to be "interacting" with the other particle. However, when an electron emits a virtual photon and then re-absorbs it, the electron is said to be interacting with itself. Self-

---

*When the muon was discovered in 1936, it looked like Yukawa's predicted particle. Gradually, however, it became evident that the muon's properties were not those of the particle in Yukawa's theory. Another eleven years passed before Yukawa's theory was confirmed.

interaction makes the world of subatomic particles a kaleidoscopic reality whose very constituents are themselves unceasing processes of transformation.

Protons, like electrons, can interact with themselves in more ways than one. The simplest proton self-interaction is the emission and re-absorption, within the time permitted by the uncertainty principle, of a virtual pion. This interaction is analogous to an electron emitting and re-absorbing a virtual photon. First there is a proton, then there is a proton and a neutral pion, then there is a proton again. Below is a Feynman diagram of a proton emitting and re-absorbing a virtual neutral pion.

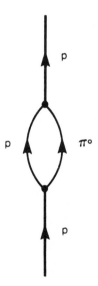

Because all protons are identical, we can assume that the original proton suddenly ceases to exist and that, at the same point in space and time, another proton and a neutral pion just as abruptly come into existence. The new proton and the neutral pion constitute a violation of the conservation law of mass-energy since their mass together is greater than the mass of the original proton. Something (the neutral pion) literally has been created out of nothing and quickly disappears again (making this a virtual process). The life span of the new particles is limited to the time calculated via the Heisenberg uncertainty principle. They quickly merge, annihilating each other, and create another proton. One blink of an eye, figuratively speaking, and the whole thing is over.

There is another way in which a proton can interact with itself. In addition to emitting and re-absorbing a neutral pion, a proton can emit a positive pion. However, by emitting a positive pion, the proton momentarily transforms itself into a neutron! First there is a proton, then there is a *neutron* (which by itself has more mass than the original proton) plus a positive pion, then there is a proton again. In other words, one of the dances that a proton does continually changes it into a neutron and back into a proton again. Below is a Feynman diagram of this dance.

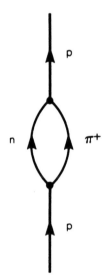

Every nucleon is surrounded by a cloud of virtual pions, which it constantly emits and re-absorbs. If a proton comes close enough to a neutron so that their virtual-pion clouds overlap, some of the virtual pions emitted by the proton are absorbed by the neutron. On the next page is a Feynman diagram of a virtual-pion exchange between a proton and a neutron.

In the left half of the diagram, a proton emits a positively charged pion, momentarily transforming itself into a neutron. Before the pion can be re-absorbed, however, it is captured by a nearby neutron. This pion capture causes the neutron to transform itself into a proton. The exchange of the positive pion causes the proton to become a neutron and the neutron to become a proton. The two original nucleons, now bound together by this exchange, have changed roles.

This is the basic Yukawa interaction. The strong force, as Yukawa theorized in 1935, is the multiple exchange of virtual pions between nucleons. The number of the exchanges (the strength of the force) increases at close range and decreases at a distance.

In a similar manner, neutrons never sit still and just be neutrons. Like protons and electrons they also constantly interact with themselves by emitting and re-absorbing virtual particles. Like protons, neutrons emit and re-absorb neutral pions. On the top of the next page is a Feynman diagram of a neutron emitting and re-absorbing a neutral pion.

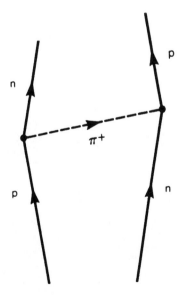

In addition to emitting a neutral pion, a neutron also can emit a negative pion. However, when a neutron emits a negative pion, it

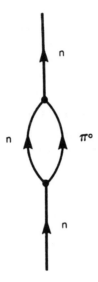

momentarily transforms itself into a proton! First there is neutron, then there is a *proton* plus a negative pion, then there is a neutron again. Below is a Feynman diagram of this dance which continually changes a neutron into a proton and back into a neutron again.

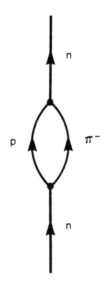

If a neutron comes so close to a proton that their virtual pion clouds overlap, some of the pions that are emitted by the neutron are absorbed by the proton. Below is a Feynman diagram of a virtual-pion exchange between a neutron and a proton.

This is another strong-force interaction. In the left half of the diagram, a neutron emits a negative pion, temporarily transforming itself into a proton. However, before the negative pion can be re-absorbed, it is captured by a nearby proton which, in turn, becomes a neutron. The exchange of a negative pion causes a neutron to become a proton and a proton to become a neutron. As before, a pair of nucleons, bound together by a virtual-pion exchange, have changed roles.

There are many more strong-force interactions. Although pions are the particles most often exchanged in the creation of the strong force, the other mesons (such as kaons, eta particles, etc.) are exchanged as well. There is no "strong force"; there are only a varying number of virtual-particle exchanges between nucleons.

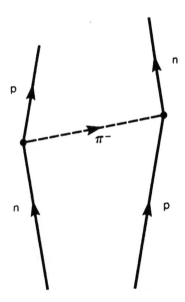

The universe, according to physicists, is held together by four fundamental types of glue. In addition to the strong force and the electromagnetic force, there is the "weak" force and the gravitational force.*

Gravity is the long-range force which holds together solar systems, galaxies, and universes. However, on the subatomic level its effect is so negligible that it is ignored altogether. Future theories, hopefully, will be able to take it into account.†

The weak force is the least known of the four forces. Its existence was inferred from the times required by certain subatomic interactions. The strong force is so short-range and powerful that strong-force interactions happen very, very fast, in about .00000000000000000000001 (10^{-23}) seconds. However, physicists discovered that a certain other type of particle interaction which they knew involved neither the electromagnetic nor the gravitational forces, required a much longer time, about .0000000001 (10^{-10}) seconds. They therefore deduced from this strange phenomenon that there must exist a fourth type of force. Since this new fourth force was known to be weaker than the electromagnetic force, it was called the weak force.

In the order of their strength, the four forces are:

> Strong (nuclear) force
> Electromagnetic force
> Weak force
> Gravity

Since the strong force and the electromagnetic force can be explained in terms of virtual particles, physicists assume that the same is true of the weak force and gravity. The particle associated with gravity is the graviton, whose properties have been theorized, but whose

---

*Recent evidence gives growing credence to the Weinberg-Salam theory that electromagnetic and weak forces are actually different manifestations of the same force, operating at different distances between particles.

†E.g., supergravity theories that use both spin 2 and spin 3/2 virtual exchange particles.

existence never has been confirmed. The particle associated with the weak force is the "W" particle, about which much has been theorized, but not much has been discovered.

The range of the strong force, relative to the electromagnetic force, is limited because mesons, relative to photons, have so much mass. Remember that the policeman who enforces the conservation law of mass-energy is willing to turn his back if the violation is quick enough, but the more flagrant the violation, the more quickly it must happen. The momentary creation of a meson out of nothing is a much more flagrant violation of the conservation law of mass-energy than the momentary creation of a photon out of nothing. Therefore, the creation and re-absorption of a meson must happen more quickly to stay within the protection, so to speak, of the uncertainty relation between time and energy. Because the life span of a virtual meson is limited, its range also is limited. The rule of thumb governing this phenomenon is this: The stronger the force, the more massive is the mediating particle, and the shorter is its range. The range of the strong force is only about one ten-trillionth (10^{-13}) of a centimeter. Accordingly, the range of the electromagnetic force is much greater than the range of the strong force. In fact, the range of the electromagnetic force is *infinite*. This is because photons *don't have any rest mass!*

"Wait a minute," says Jim de Wit, agreeing with us for a change. "This doesn't make sense. A virtual photon is a photon which is emitted and re-absorbed quickly enough to avoid violating the conservation law of mass-energy. Right?"

"Right," says a particle physicist on his way to the cyclotron.

"Then how can a particle, or anything else, be emitted and re-absorbed within certain time limits, like the time limits imposed by the uncertainty principle, and still have an *infinite* range? It doesn't make sense."

De Wit has a point. At first glance it appears that he is correct. On closer examination, however, there *is* a subtle logic involved which *does* make sense. If the limitations of the conservation law of mass-energy are avoided by a balance of time and energy (mass) permitted by the uncertainty principle, and a virtual photon has *no* (rest) mass, then it has all the time in the world, literally, to go where it pleases. In other words, there is no practical difference between a "real" photon and a "virtual" photon. The only difference between them is that the creation of a "real" photon does not violate the conservation law of mass-energy and the creation of a "virtual" photon avoids the law momentarily via the Heisenberg uncertainty principle.

This is a good example of how "unreal" and "ivory-tower-like" the nonmathematical explanation of a successful physical theory can sound. The reason for this is that physical theories, in order to describe more accurately the phenomena under consideration, have become more and more divorced from everyday experience (i.e., more abstract). Although these highly abstract theories, such as quantum theory and relativity, are unaccountably accurate to an awesome degree, they truly are "free creations" of the human mind. Their primary link with ordinary experience is not the abstract content of their formalisms, but the fact that, somehow, they work.*

The distinction between a transient, virtual (nothing-something-nothing) state and a "real" one (something-something-something) is similar to the Buddhist distinction between reality as it actually is and the way that we usually see it. For example, Feynman himself described the difference between a virtual state and a real state (of a photon) as a matter of perspective.

> . . . what looks like a real process from one point of view may appear as a virtual process occurring over a more extended time.
>
> For example, if we wish to study a given real process, such as the scattering of light, we can, if we wish, include in principle

*Paul Schilpp (ed.), *Albert Einstein, Philosopher-Scientist*, vol. 1, New York, Harper & Row, 1949, has some good essays on this theme.

the source, scatterer, and eventual absorber of the scattered light in our analysis. We may imagine that no photon is present initially, and that the source then emits light. . . . The light is then scattered and eventually absorbed. . . . From this point of view the process is virtual, that is, we start with no photons and end with none. Thus we can analyze the process by means of our formulas for real processes by attempting to break the analysis into parts corresponding to emission, scattering, and absorption.[2]

According to Buddhist theory, reality is "virtual" in nature. What appear to be "real" objects in it, like trees and people, actually are transient illusions which result from a limited mode of awareness. The illusion is that parts of an overall virtual process are "real" (permanent) "things." "Enlightenment" is the experience that "things," including "I," are transient, virtual states devoid of separate existences, momentary links between illusions of the past and illusions of the future unfolding in the illusion of time.

Particle self-interactions become quite intricate when virtual particles emit virtual particles which emit virtual particles in a diminishing sequence. On the next page is a Feynman diagram of a virtual particle (a negative pion) transforming itself momentarily into two more virtual particles, a neutron and an anti-proton (Dirac's 1928 theory also predicted anti-protons which were discovered at Berkeley in 1955).

This is the simplest example of self-interaction. On page 265 is the exquisite dance of a single proton performed in the flicker of time permitted by the uncertainty principle. This diagram was constructed by Kenneth Ford in his book, *The World of Elementary Particles*.[3] Eleven particles make their transient appearance between the time the original proton transforms itself into a neutron and a pion and the time it becomes a single proton again.

A proton never remains a simple proton. It alternates between being a proton and a neutral pion on the other hand, and being a

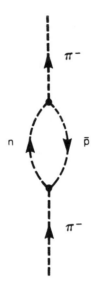

neutron and a neutral pion on the other hand. A neutron never remains a simple neutron. It alternates between being a neutron and a neutral pion on the one hand, and being a proton and a negative pion on the other hand. A negative pion never remains a simple negative pion. It alternates between being a neutron and an anti-proton on the one hand, etc., etc. In other words, *all particles exist potentially (with a certain probability) as different combinations of other particles.* Each combination has a certain probability of happening.

Quantum theory deals with probability. The probability of each of these combinations can be calculated with accuracy. According to quantum theory, however, it is ultimately chance that determines which of these combinations actually occur.

The quantum view that all particles exist potentially as different combinations of other particles parallels a Buddhist view, again. According to *The Flower Garland Sutra*, each part of physical reality is constructed of all the other parts. (A *sutra* is a written account of the Buddha's teachings.) This theme is illustrated in *The Flower Garland Sutra* by the metaphor of Indra's net. Indra's net is a vast network of gems which overhangs the palace of the god Indra.

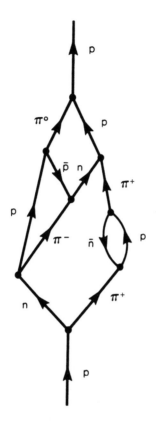

In the words of an English interpreter:

In the heaven of Indra, there is said to be a network of pearls, so arranged that if you look at one you see all the others reflected in it. In the same way each object in the world is not merely itself but involves every other object and in fact *is* everything else.[4]

The appearance of physical reality, according to Mahayana Buddhism, is based upon the interdependence of all things.*,†

Although this book is not about physics and Buddhism specifically, the similarities between the two, especially in the field of particle physics, are so striking and plentiful that a student of one necessarily must find value in the other.

Now we come to the most psychedelic aspect of particle physics. Below is a Feynman diagram of a three-particle interaction.

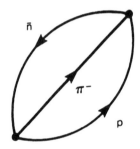

In this diagram no world line leads up to the interaction and no world line leads away from it. It just happens. It happens literally out of nowhere, for no apparent reason, and without any apparent cause. Where there was *no-thing*, suddenly, in a flash of spontaneous existence, there are three particles which vanish without a trace.

*From discussions with Prof. John Blofeld, a Buddhist and Taoist scholar, I believe that there are even better illustrations of this concept in *The Flower Garland Sutra* than the metaphor of Indra's net. (*The Flower Garland Sutra*, which also is called the *Hua Yen Sutra* [Chinese] and the *Avatamsaka Sutra* [Sanskrit], is extremely long. A complete translation with commentary would be about 150 volumes.) At the time of this printing there is no complete English translation of *The Flower Garland Sutra*, although one is in progress by the Buddhist Text Translation Society, City of Ten Thousand Buddhas, Talmage, California 95481.

†G. F. Chew's bootstrap theory may be a physical analog to the Buddhist theory of interdependent originations.

This type of Feynman diagram is called a "vacuum diagram.*
That is because the interactions happen in a vacuum. A "vacuum," as
we normally construe it, is a space that is entirely empty. Vacuum
diagrams, however, graphically demonstrate that there is no such
thing. From "empty space" comes something, and then that something disappears again into "empty space."

In the subatomic realm, a vacuum obviously is *not* empty. So
where did the notion of a completely empty, barren, and sterile
"space" come from? We made it up! There is no such thing in the real
world as "empty space." It is a mental construction, an idealization,
which we have taken to be true.

"Empty" and "full" are "false distinctions" that *we have created,*
like the distinction between "something" and "nothing." They are
abstractions from experience which we have mistaken for experience.
Perhaps we have lived so long in our abstractions that instead of realiz-
ing that they are drawn from the real world we believe that they *are*
the real world.

Vacuum diagrams are the serious product of a well-intentioned
physical science. However, they also are wonderful reminders that we
can intellectually create our "reality." It is not possible, according to
our usual conceptions, for "something" to come out of "empty
space"; but, at the subatomic level, *it does,* which is what vacuum
diagrams illustrate. In other words, there is no such thing as "empty
space" (or "nothing") except as a concept in our categorizing minds.

The core *sutras* of Mahayana Buddhism (the type of Buddhism
practiced in Tibet, China, and Japan) are called the *Prajnaparamita
Sutras.*† Among the most central of the *Prajnaparamita Sutras* (there

---

*Brian Josephson, Jack Sarfatti, and Nick Herbert independently have spec-
ulated that human sensory systems might detect the zero-point vacuum fluctuations of the dance of virtual particles in empty space predicted by the uncertainty
principle. If this is so, such detections might be part of the mechanism of mystic
knowing.

†*Prajna* (Sanskrit) means "wisdom," but it is a special kind of wisdom
which cannot be learned through studying books. *Paramita* (literally "to cross
over") means "bringing something to perfection."

are twelve volumes of them) is a *sutra* which is called simply, *The Heart Sutra*. *The Heart Sutra* contains one of the most important ideas of Mahayana Buddhism:

   . . . form is emptiness, emptiness is form.

   Below is a vacuum diagram of six different mutually interacting particles.

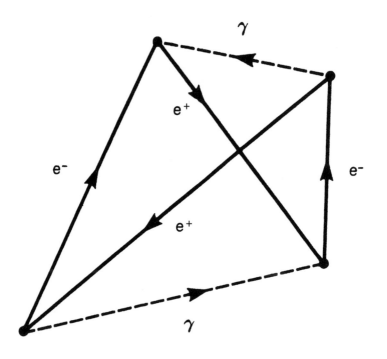

It depicts an exquisite dance of emptiness becoming form and form becoming emptiness. Perhaps, as the wise people of the East have written, form *is* emptiness and emptiness *is* form.

   In any case, vacuum diagrams are representations of remarkable transformations of "something" into "nothing" and "nothing" into "something." These transformations occur continuously in the subatomic realm and are limited only by the uncertainty principle, the conservation laws, and probability.*

There are roughly twelve conservation laws. Some of them affect every type of subatomic interaction. Some of them affect only some types of subatomic interaction. There is a simple rule of thumb to remember: The stronger the force, the more its interactions are restrained by conservation laws. For example, strong interactions are restrained by all twelve conservation laws; electromagnetic interactions are restrained by eleven of the conservation laws; and weak interactions are restrained by only eight of the conservation laws.† Gravitational interactions, those involving the most feeble force in the subatomic world, have not been studied yet (no one has found a graviton), but they may violate even more conservation laws.

Nonetheless, where the conservation laws have jurisdiction, they are inviolable rules which shape the form of all particle interactions. For example, the conservation law of mass-energy dictates that all spontaneous particle decays be "downhill." When a single particle spontaneously decays, it always decays into lighter particles. The total mass of the new particles is always less than the mass of the original particle. The difference between the mass of the original particle and

---

*Conservation laws impose absolute checks, but the probability laws can effectively exclude much of what the conservation laws would permit. They impose a great deal of structure.

†The strong interactions are restrained by all twelve conservation laws: energy, momentum, angular momentum, charge, electron-family number, muon-family number, baryon-family number, time reversal (T), combined space inversion and charge conjugation (PC), space inversion alone (P) and charge conjugation alone (C), strangeness, and isotopic spin.

Electromagnetic interactions, next down on the ladder of strengths, lose isotopic spin conservation. Weak interactions, one rung lower, lose strangeness conservation, parity conservation, and charge conjugation invariance as well (but the combination PC remains valid). The last step down the ladder to gravitational interactions at the microscale has not been taken.

the total mass of the new particles is converted into the kinetic energy of the new particles (which fly away).

"Uphill" interactions are only possible when kinetic energy, in addition to the energy of being (mass) of the original particles, is available for the creation of new particles. Two colliding protons, for example, can create a proton, a neutron, and a positive pion. The total mass of these new particles is greater than the mass of the two original protons. This is possible because some of the kinetic energy of the projectile proton went into the creation of the new particles.

In addition to mass-energy, momentum is conserved in every particle interaction. The total momentum carried by particles going into an interaction must equal the total momentum of the particles leaving the interaction. This is why the spontaneous decay of a single particle always produces at least two new particles. A particle at rest has zero momentum. If it decays into a single new particle which then flies off, the momentum of the new particle will exceed the momentum of the original particle (zero). The momenta of at least two new particles flying off in opposite directions, however, cancel each other, producing a total momentum of zero.

Charge also is conserved in every particle interaction. If the total charge of the particles entering an interaction is plus two (for example, two protons), the total charge of the particles leaving the interaction must also equal plus two (after the positive and negative particles cancel each other). Spin, too, is conserved, although keeping the books balanced in regard to spin is more complicated than it is in regard to charge.

In addition to the conservation laws of mass-energy, momentum, charge, and spin, there are conservation laws of family numbers. For example, if two baryons, or heavy-weight particles (like two protons), go into an interaction, two baryons must be among the resulting new particles (like a neutron and a lambda particle).

This same baryon conservation law, along with the conservation law of mass-energy, "explains" why protons are stable particles (i.e., why they do not decay spontaneously). Spontaneous decays must be downhill to satisfy the conservation law of mass-energy. Protons cannot decay downhill without violating the conservation law of baryon family numbers because protons are the lightest baryons. If a proton were to decay spontaneously, it would have to decay into particles lighter than itself, but there are no baryons lighter than a proton. In other words, if a proton were to decay, there would be one less baryon in the world. In fact, this never happens. This scheme (the conservation law of baryon family numbers) is the only way that physicists so far have been able to account for the proton's stability. A similar conservation law of lepton family numbers accounts for the stability of electrons. (There are no lighter leptons than an electron.)

Some of the twelve conservation laws are actually "invariance principles." An invariance principle is a law that says, "under a change of circumstances (like changing the location of an experiment) all of the laws of physics remain valid." "All of the laws of physics," so to speak, is the "conserved quantity" of an invariance principle. For example, there is a time-reversal invariance principle. In order for a process to be possible, according to this principle, it must be reversible in time. If a positron-electron annihilation can create two photons (it can), then the annihilation of two photons can create a positron and an electron (it can).

Conservation laws and invariance principles are based on what physicists call symmetries. The fact that space is the same in all directions (isotropic) and in all places (homogeneous) is an example of symmetry. The fact that time is homogeneous is another example. These symmetries simply mean that a physics experiment performed in Boston this spring will give the same result as the same experiment performed in Moscow next fall.

In other words, physicists now believe that the most fundamental laws of physics, the conservation laws and invariance principles, are based upon those foundations of our reality that are so basic that they go unnoticed. This does not mean (probably) that it has taken physicists three hundred years to realize that moving an object, like a telephone, around the country does not distort its shape or size (space is homogeneous), nor does turning it upside down (space is isotropic), nor does letting it get two weeks older (time is homogeneous). Everyone knows that this is the way our physical world is constructed.

Where and when a subatomic experiment is performed are not critical data. The laws of physics do not change with time and place.

It does mean, however, that it has taken physicists three hundred years to realize that the most simple and beautiful mathematical structures may be those that are based on these unobtrusively obvious conditions.

Theoretical physics, roughly speaking, has branched into two schools. One school follows the old way of thinking and the other school follows new ways of thinking. Physicists who follow the old way of thinking continue their search for the elementary building blocks of the universe in spite of the hall-of-mirrors predicament (page 215).

For these physicists, the most likely candidate at present for the title of "ultimate building block of the universe" is the quark. A quark is a type of hypothetical particle theorized by Murray Gell-Mann in 1964. It is named after a word in James Joyce's book *Finnegans Wake*.

All known particles, the theory goes, are composed of various combinations of a few (twelve) different types of quarks. The great quark hunt could become very exciting, but no matter what is discovered, one thing about it already is certain: The discovery of quarks will open an entirely new area of research, namely, "What are *quarks* made of?"

The physicists who follow the new ways of thinking are pursuing so many different approaches to understanding subatomic phenomena that it is not possible to present them all. Some of these physicists feel that space and time are all that there is. According to this theory, actors, action, and stage are all manifestations of an underlying four-dimensional geometry. Others (like David Finkelstein) are exploring processes which lie "beneath time," processes *from which* space and time, the very fabric of experiential reality, are derived. These theories, at the moment, are speculative. They cannot be "proven" (demonstrated mathematically).

The most successful departure from the unending search-for-the-ultimate-particle syndrome is the S-Matrix theory. In S-Matrix theory, the dance rather than the dancers is of primary importance. S-Matrix theory is different because it places the emphasis upon interactions rather than upon particles.

"S Matrix" is short for Scattering Matrix. Scattering is what happens to particles when they collide. A matrix is a type of mathematical table. An S Matrix is a table of probabilities (page 121).

When subatomic particles collide several things usually are possible. For example, the collision of two protons can create (1) a proton, a neutron, and a positive pion, (2) a proton, a lambda particle, and a positive kaon, (3) two protons and six assorted pions, (4) numerous other combinations of subatomic particles. Each of these possible combinations (which are the combinations that do not violate the conservation laws) occurs with a certain probability. In other words, some of them occur more often than others. The probabilities of various combinations in turn depend upon such things as how much momentum is carried into the collision area.

In an S Matrix all of these probabilities are tabulated in such a way that we can look up or calculate the possible results of any collision along with their probabilities if we know what particles initially collide and how much momentum they have. Of course, there are so many possible combinations of particles (each one of which can yield a variety of results) that a complete matrix (table) containing all the probabilities of all the possible combinations of particles would be enormous. In fact, such a complete table has not been compiled. This is no immediate problem, however, since physicists are concerned only with a small part of the S Matrix at any one time (for example, the part which deals with two-proton collisions). Such parts of the total S Matrix are called elements of the S Matrix. The major limitation of S-Matrix theory is that at present it applies only to the strongly interacting particles (mesons and baryons), which, as a group, are called hadrons (hay'drons).

On the next page is an S-Matrix diagram of a subatomic interaction. It is very simple. The collision area is the circle. Particles 1 and 2 go into the collision area and particles 3 and 4 come out of the collision area. The diagram tells nothing about what happened at the point of collision. It shows only what particles went into the interaction and what particles came out of the interaction.

An S-Matrix diagram is not a space-time diagram. It does not show the position of the particles in space or time. This is intentional because we do not know the exact positions of the interacting particles. We have chosen to measure their momenta and consequently their position is unknown (the Heisenberg uncertainty principle). For

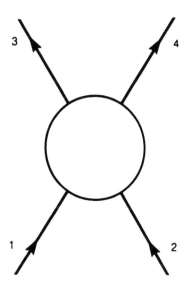

this reason, S-Matrix diagrams indicate only that an interaction took place in a certain area (inside the circle). They are purely symbolic representations of particle interactions.

Not all interactions involve only two initial particles and two final particles. Below are some other forms that an S-Matrix diagram can take.

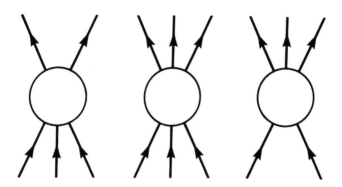

Like Feynman diagrams, S-Matrix diagrams can be rotated. The direction of the arrowheads distinguishes the particles from the antiparticles. Here is an S-Matrix diagram of a proton colliding with a negative pion to produce a proton and a negative pion.

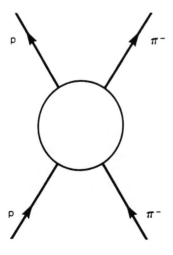

When this diagram is rotated it becomes a diagram of a proton/anti-proton annihilation producing a negative pion and a positive pion. (The positive pion is the anti-particle of the negative pion in the original reaction.)

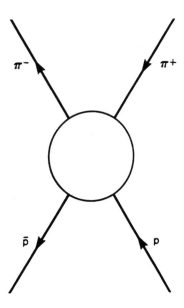

Every time a diagram is rotated it depicts another possible interaction. This particular S-Matrix diagram can be rotated four times. All of the particles that can be depicted by rotating a single element of the S Matrix are intimately related to each other. In fact, *all of the particles represented in an S-Matrix diagram (including those discovered by rotating the diagram) are defined in terms of each other.* Which of them are "elementary" is a meaningless question.

Since the particles resulting from an interaction often become involved in other interactions, separate elements of the S Matrix can be assembled diagrammatically into a network of related interactions. Each network, as well as each interaction, is associated with a certain probability. These probabilities can be calculated.

According to S-Matrix theory, "particles" are *intermediate states* in a network of interactions. The lines in an S-Matrix diagram are not the world lines of different particles. Lines in an S-Matrix diagram of an interaction network are "reaction channels" through which energy flows. A "neutron," for example, is a *reaction channel*. It can be formed by a proton and a negative pion.

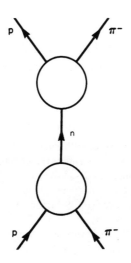

If more energy is available, however, the same channel can be created by a lambda particle and a neutral kaon, and several other particle combinations as well.

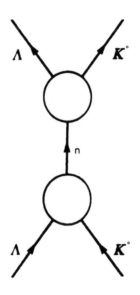

In short, S-Matrix theory is based upon *events,* not upon things.*,† Dancers no longer stand apart as significant entities. In fact, the dancers are not even defined except in terms of each other. In S-Matrix theory there is only the dance.

---

*S-Matrix theory is concerned with events in the sense of overall results of a process rather than in the sense of individual things happening during the collision process. There are well-defined entities in the input and output channels (or the S Matrix would not be defined), but in the interaction region itself (inside the circle) everything is blurred and unspecified. "S-Matrix philosophy," according to Brian Josephson at Cambridge University, "is a statement of unanalyzability in detail."

†A lay discussion of S-Matrix theory is contained in *The Tao of Physics* by F. Capra, Berkeley, Shambhala, 1975, pp. 261–76.

*   *   *

We have come a long way from Newton and his proverbial apple. Nonetheless, apples are a real part of the apparent world. When we eat an apple we are aware of who is eating and what is being eaten as distinct from the action of eating.

The idea that objects exist apart from event is part of the epistemological net with which we snare our particular form of experience. This idea is dear to us because we have accepted it, without question, as the basis of our reality. It profoundly influences how we see ourselves. It is the root of our inescapable sense of separateness from others and environment.

The history of scientific thought, if it teaches us anything at all, teaches us the folly of clutching ideas too closely. To this extent it is an echo of eastern wisdom which teaches us the folly of clutching anything.

Part One

ENLIGHTENMENT

1

More Than Both

What does physics have in common with enlightenment? Physics and enlightenment apparently belong to two realms which are forever separate. One of them (physics) belongs to the external world of physical phenomena and the other of them (enlightenment) belongs to the internal world of perceptions. A closer examination, however, reveals that physics and enlightenment are not so incongruous as we might think. First, there is the fact that only through our perceptions can we observe physical phenomena. In addition to this obvious bridge, however, there are more intrinsic similarities.

Enlightenment entails casting off the bonds of concept ("veils of ignorance") in order to perceive directly the inexpressible nature of undifferentiated reality. "Undifferentiated reality" is the same reality that we are a part of now, and always have been a part of, and always will be a part of. The difference is that we do not look at it in the same way as an enlightened being. As everyone knows(?), words only *represent (re-present)* something else. They are not real things. They are only *symbols.* According to the philosophy of enlightenment, *everything* (every*thing*) is a symbol. The reality of symbols is an illusory reality. Nonetheless, it is the one in which we live.

Although undifferentiated reality is inexpressible, we can talk around it (using more symbols). The physical world, as it appears to the unenlightened, consists of many separate parts. These separate parts, however, are not really separate. According to mystics from around the world, each moment of enlightenment (grace/insight/samadhi/satori) reveals that everything—all the separate parts of the universe—are manifestations of the same whole. There is only *one* reality, and it is whole and unified. It is one.

We already have learned that understanding quantum physics requires a modification of ordinary conceptions (like the idea that something cannot be a wave *and* a particle). Now we shall see that physics may require a more complete alteration of our thought processes than we ever conceived or, in fact, than we ever could conceive. Likewise we previously have seen that quantum phenomena seem to make decisions, to "know" what is happening elsewhere (page 69). Now we shall see how quantum phenomena may be connected so intimately that things once dismissed as "occult" could become topics of serious consideration among physicists.

In short, both in the need to cast off ordinary thought processes (and ultimately to go "beyond thought" altogether), and in the perception of reality as one unity, the phenomenon of enlightenment and the science of physics have much in common.

Enlightenment is a state of being. Like *all* states of being it is indescribable. It is a common misconception (literally) to mistake the description of a state of being for the state itself. For example, try to describe happiness. It is impossible. We can talk around it, we can describe the perspectives and actions that usually accompany a state of happiness, but we cannot describe happiness itself. Happiness and the description of happiness are two different things.

Happiness is a state of being. That means that it exists in the realm of direct experience. It is the intimate perception of emotions and sensations which, indescribable in themselves, constitute the state of happiness. The word "happiness" is the label, or symbol, which we pin on this indescribable state. "Happiness" belongs to the realm of abstractions, or concepts. A state of being is an *experience*. A description of a state of being is a *symbol. Symbols and experience do not follow the same rules.*

This discovery, that symbols and experience do not follow the same rules, has come to the science of physics under the formidable title of quantum logic. The possibility that separate parts of reality (like you and I and tugboats) may be connected in ways which both our common experience and the laws of physics belie, has found its way into physics under the name of Bell's theorem. Bell's theorem and quantum logic take us to the farthest edges of theoretical physics. Many physicists have not even heard of them.

Bell's theorem and quantum logic (currently) are unrelated. Proponents of one seldom are interested in the other. Nonetheless, they have much in common. They are what is really *new* in physics. Of course, laser fusion (fusing atoms with high-energy light beams) and the search for quarks generally are considered to be the frontiers of theoretical physics.* In a certain sense, they are. However, there is a big difference between these projects and Bell's theorem and quantum logic.

Laser fusion research and the great quark hunt are endeavors within the existing paradigms of physics. A paradigm is an established thought process, a framework. Both quantum logic and Bell's theorem are potentially explosive in terms of existing frameworks. The first (quantum logic) calls us back from the realm of symbols to the realm of experience. The second (Bell's theorem) tells us that there is no such thing as "separate parts." All of the "parts" of the universe are connected in an intimate and immediate way previously claimed only by mystics and other scientifically objectionable people.

The central mathematical element in quantum theory, the hero of the story, is the wave function. The wave function is that mathematical

---

*Laser fusion and the search for quarks already have become the partial domain of experimental physics. The new frontiers of *theoretical* physics appear to be solitons and unified gauge theories.

entity which allows us to determine the possible results of an interaction between an observed system and an observing system. The celebrated position held by the wave function is due not only to Erwin Schrödinger, who discovered it, but also to the Hungarian mathematician John von Neumann.

In 1932, von Neumann published a famous mathematical analysis of quantum theory called *The Mathematical Foundations of Quantum Mechanics.*[1] In this book von Neumann, in effect, asked the question, "If a 'wave function,' this purely abstract mathematical creation, actually should describe something in the real world, what would that something be like?" The answer that he deduced is exactly the description of a wave function that we already have discussed (page 80).

This strange animal constantly would change with the passage of time. Each moment it would be different than the moment before. It would be a composite of all the possibilities of the observed system which it describes. It would not be a simple mixture of possibilities, it would be a sort of organic whole whose parts are changing constantly but which, nonetheless, is somehow a thing-in-itself.

This thing-in-itself would continue to develop indefinitely until an observation (measurement) is made on the observed system which it represents. If the "observed system" is a photon "propagating in isolation," the wave function representing this photon would contain all of the possible results of the photon's interaction with a measuring device, like a photographic plate.* (For example, the possibilities contained in the wave function might be that the photon will be detected in area A of the photographic plate, that the photon will be detected in area B of the photographic plate, and that the photon will be detected in area C of the photographic plate.)

Once the photon is set in motion the wave function associated with it would continue to develop (change) according to a causal law

---

*There are several interpretations of the formalism of quantum mechanics. Von Neumann thought that only ensembles, i.e., groups of photons, have wave functions and not single particles. A few physicists still agree with this point of view, although most physicists do not.

(the Schrödinger wave equation) until the photon interacts with the observing system. At that instant, one of the possibilities contained in the wave function would actualize and the other possibilities contained in the wave function would cease to exist. They simply would disappear. The wave function, that strange animal that von Neumann was attempting to describe, would "collapse." The collapse of this particular wave function would mean that the probability of one of the possible results of the photon-measuring-device interaction became *one* (it happened) and the probability of the other possibilities became *zero* (they were no longer possible). After all, a photon can be detected only in one place at a time.

The wave function, according to this view, is not quite a thing yet it is more than an idea. It occupies that strange middle ground between idea and reality, where all things are possible but none are actual. Heisenberg likened it to Aristotle's *potentia* (page 72).

This approach has unconsciously shaped the language, and therefore the thinking, of most physicists, even those who consider the wave function to be a mathematical fiction, an abstract creation whose manipulation somehow yields the probabilities of real events which happen in real (versus mathematical) space and time.

Needless to say, this approach also has caused a great deal of confusion, which is as unclear today as it was in von Neumann's time. For example, exactly when does the wave function collapse? (The Problem of Measurement, page 87.) Is it when the photon strikes the photographic plate? Is it when the photographic plate is developed? Is it when we look at the developed plate? Exactly *what* is it that collapses? Where does the wave function live before it collapses? and so on. This view of the wave function, that it can be described as a real thing, is generally the view of the wave function attributed to von Neumann. However, the real-wave-function description is only one of two approaches to understanding quantum phenomena which he discussed in *The Mathematical Foundations of Quantum Mechanics*.

The second approach, to which von Neumann devoted much less time, is a re-examination of the language by which it is necessary to express quantum phenomena. In the section "Projections as Propositions," he wrote:

> . . . the relation between the properties of a physical system on the one hand, and the projections [wave function] on the other, makes possible a sort of logical calculus with these. However, in contrast to the concepts of ordinary logic, this system is extended by the concepts of "simultaneous decidability" [the uncertainty principle] which is characteristic for quantum mechanics.[2]

This suggestion, that the novel properties of quantum theory can be used to construct a "logical calculus" which is "in contrast to the concepts of ordinary logic," is what von Neumann considered the alternative to describing wave functions as real things.

Most physicists, however, have adopted a third explanation of wave functions. They dismiss them as purely mathematical constructions, abstract fictions which represent nothing in the world of reality. Unfortunately, this explanation leaves forever unanswered the question, "How, then, can wave functions predict so accurately probabilities which can be verified through actual experience?" In fact, how can wave functions predict *anything* when they are defined as completely unrelated to physical reality. This is a scientific version of the philosophical question, "How can mind influence matter?"

Von Neumann's second approach to understanding the paradoxical puzzles of quantum phenomena took him far beyond the boundaries of physics. This brief work pointed to a fusion of ontology, epistemology, and psychology which only now is beginning to emerge. In short, the problem, said von Neumann, is in the language. Herein lies the germ of what was to become quantum logic.

In pointing to the problem of language, von Neumann put his finger on why it is so difficult to answer the question, "What is quantum mechanics?" Mechanics is the study of motion. Therefore, quantum mechanics is the study of the motion of quanta—but what are quanta? According to the dictionary, a quantum is a quantity of something. The question is, a quantity of what?

A quantum is a piece of action (a piece of the action?). The problem is that a quantum can be like a wave, and then again it can be like a particle, which is everything that a wave isn't. Furthermore, when a quantum is like a particle, it is not like a particle in the ordinary sense of the word. A subatomic "particle" is not a "thing." (We cannot determine simultaneously its position and momentum.) A subatomic "particle" (quantum) is a set of relationships, or an intermediate state. It can be broken up, but out of the breaking come more particles as elementary as the original. ". . . Those who are not shocked when they first come across quantum theory," said Niels Bohr, "cannot possibly have understood it."[3]

Quantum theory is not difficult to explain because it is complicated. Quantum theory is difficult to explain because the words which we must use to communicate it are not adequate for explaining quantum phenomena. This was well known and much discussed by the founders of quantum theory. Max Born, for example, wrote:

> The ultimate origin of the difficulty lies in the fact (or philosophical principle) that we are compelled to use words of common language when we wish to describe a phenomenon, not by logical or mathematical analysis, but by a picture appealing to the imagination. Common language has grown by everyday experience and can never surpass these limits. Classical physics has restricted itself to the use of concepts of this kind; by analyzing visible motions it has developed two ways of representing them by elementary processes: moving particles and waves. There is no other way of giving a pictorial description of motions—we have to apply it even in the region of atomic process, where classical physics break down.[4]

This is the view currently held by most physicists: We encounter problems explaining subatomic phenomena when we try to visualize them. Therefore, it is necessary to forgo explanations in terms of "common language" and restrict ourselves to "mathematical analysis." To learn the physics of subatomic phenomena we first must learn mathematics.

"Not so!" says David Finkelstein, Director of the School of Physics at the Georgia Institute of Technology. Mathematics, like English, also is a language. It is constructed of symbols. "The best you can get with symbols is a maximal but incomplete description."[5] A mathematical analysis of subatomic phenomena is no better qualitatively than any other symbolic analysis, because *symbols do not follow the same rules as experience*. They follow rules of their own. In short, the problem is not *in* the language, the problem *is* the language.

The difference between experience and symbol is the difference between mythos and logos. Logos imitates, but can never replace, experience. It is a *substitute* for experience. Logos is the artificial construction of dead symbols which mimics experience on a one-to-one basis. Classical physical theory is an example of a one-to-one correspondence between theory and reality.

Einstein argued that no physical theory is complete unless every element in the real world has a definite counterpart in the theory. Einstein's theory of relativity is the last great classical theory (even though it is a part of the new physics) because it is structured in a one-to-one way with phenomena. Unless a physical theory has one-to-one correspondence with phenomena, argued Einstein, it is not complete.

> Whatever the meaning assigned to the term *complete*, the following requirement for a complete theory seems to be a necessary one: *every element of the physical reality must have a counterpart in the physical theory.*[6] [Italics in the original.]

Quantum theory does not have this one-to-one correspondence between theory and reality (it cannot predict individual events—only probabilities). According to quantum theory, individual events are chance happenings. There are no theoretical elements in quantum theory to correspond with each individual event that actually happens. Therefore quantum theory, according to Einstein, is incomplete. This was a basic issue of the famous Bohr-Einstein debates.

Mythos points toward experience, but it does not replace experience. Mythos is the opposite of intellectualism. Ceremonial chants at primitive rituals (like football games) are good examples of mythos. They endow experience with value, originality, and vitality, but they do not seek to replace it.

Theologically speaking, logos is the original sin, the eating of the fruit of knowledge, the expulsion from the Garden of Eden. Historically speaking, logos is the growth of the literary revolution, the birth of the written tradition out of the oral tradition. From any point of view, logos (literally) is a dead letter. "Knowledge," wrote e. e. cummings, "is a polite word for / dead but not buried imagination." He was talking about logos.

Our problem, according to Finkelstein, is that we cannot understand subatomic phenomena, *or any other kind of experience,* through the use of symbols alone. As Heisenberg observed:

> The concepts initially formed by abstraction from particular situations or experiential complexes acquire a *life of their own.*[7] [Italics added].

Getting lost in the interaction of symbols is analogous to mistaking the shadows on the wall of the cave for the real world outside the cave (which is direct experience). The answer to this predicament is to approach subatomic phenomena, as well as experience in general, with a language of mythos rather than a language of logos.

Finkelstein put it this way:

> If you want to envision a quantum as a dot then you are trapped. You are modeling it with classical logic. The whole point is that there *is no* classical representation for it. We have to learn to live with the experience.
>
> Question: How do you communicate the experience?
>
> Answer: You don't. But by telling how you make quanta and how you measure them, you enable others to have it.[8]

According to Finkelstein, a language of mythos, a language which alludes to experience but does not attempt to replace it or to mold our perception of it is the true language of physics. This is because not only the language that we use to communicate our daily experience, but also mathematics, follows a certain set of rules (classical logic). *Experience itself is not bound by these rules.* Experience follows a much more permissive set of rules (quantum logic). Quantum logic is not only more exciting than classical logic, it is more real. It is based not upon the way that we *think* of things, but upon the way that we *experience* them.

When we try to describe experience with classical logic (which is what we have been doing since we learned to write), we put on a set of blinders, so to speak, which not only restricts our field of vision, but also distorts it. These blinders are the set of rules known as classical logic. The rules of classical logic are well defined. They are simple. The only problem is that they do not correspond to experience.

The most important difference between the rules of classical logic and the rules of quantum logic involves the law of distributivity. The law of distributivity, or the distributive law, says that "A, and B or C" is the same as "A and B, or A and C." In other words, "I flip a coin and it comes up heads or tails" has the same meaning as "I flip a coin and it comes up heads, or I flip a coin and it comes up tails." The distributive law, which is a foundation of classical logic, *does not apply to quantum logic.* This is one of the most important but least understood aspects of von Neumann's work. In 1936, von Neumann and his colleague, Garrett Birkhoff, published a paper which laid the foundations of quantum logic.[9]

In it they used an example of a familiar (to physicists) phenomenon to disprove the distributive law. By so doing they demonstrated mathematically that it is impossible to describe experience (including subatomic phenomena) with classical logic, because the real world follows different rules. The rules that experience follows they called quantum logic. The rules which symbols follow they called classical logic.

Finkelstein uses a version of Birkhoff and von Neumann's example to disprove the law of distributivity. Finkelstein's demonstration requires only three pieces of plastic. These three pieces of plastic are contained in the envelope attached to the back cover of this book. Remove them from their envelope now and examine them.* Notice that they are transparent and tinted about the color of sunglasses. In fact, pieces of plastic just like these but thicker *are* used for sunglasses. They are very effective in reducing glare because of their particular characteristics. These pieces of plastic are called polarizers and, of course, the sunglasses which use them are called polaroid sunglasses.

Polarizers are a special kind of light filter. Most frequently they are made of stretched sheets of plastic material in which all the molecules are elongated and aligned in the same direction. Under magnification the molecules look something like this.

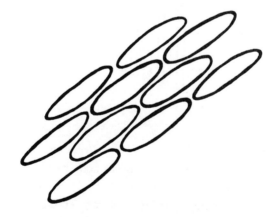

*It was necessary for me to decide between including the polarizers mentioned in the text and keeping the price of this book within the reach of every person. I chose to omit the polarizers. However, even though it is impossible for words to convey experience, I have kept the text as I originally wrote it to convey the flavor of the demonstration. (Small sheets of polarized plastic are very inexpensive and can be purchased through most popular scientific catalogues.)

These long, slender molecules are responsible for the polarization of the light which passes through them.

The polarization of light can be understood most easily as a wave phenomenon. Light waves from an ordinary light source, like the sun, emanate in every fashion, vertically, horizontally, and every way in between. This does not mean only that light radiates from a source in all directions. It means that in any given beam of light some of the light waves are vertical, some of the light waves are horizontal, some are diagonal, and so forth. To a light wave, a polarizer looks something like a picket fence. Whether it can get through the fence or not depends upon whether it is aligned with the fence or not. If the polarizer is aligned vertically, only the vertical light waves make it through. All of the other light waves are obstructed. All of the light waves that pass through a vertical polarizer are aligned vertically. This light is called vertically polarized light.

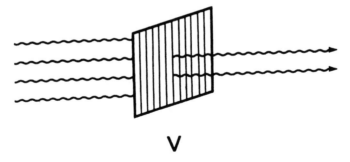

V

If the polarizer is aligned horizontally, only the horizontal light waves make it through. All of the other light waves are obstructed (first illustration, next page). All of the light waves that pass through a horizontal polarizer are aligned horizontally. This light is called horizontally polarized light.

No matter how the polarizer is aligned, all of the light waves passing through it are aligned in the same plane. The arrows on the polarizers indicate the direction in which the light passing through them is polarized (which way the molecules in the plastic are elongated).

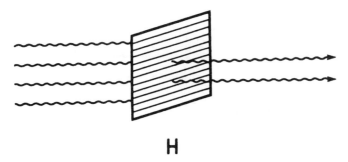

H

Take one of the polarizers and hold it with the arrow pointing up (or down). The light coming through this polarizer now is polarized vertically. Now take another polarizer and hold it behind the first polarizer with its arrow also pointing up (or down). Notice that, except for a slight attenuation due to the tint, all of the light that gets through the first polarizer also gets through the second polarizer.

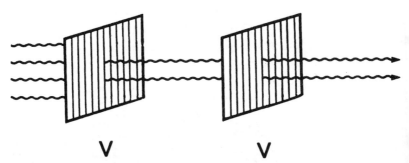

V   V

Now rotate one of the polarizers from vertical to horizontal. As it is rotated, notice that less and less light gets through the pair. When one of the polarizers is vertical and the other polarizer is horizontal, no light gets through them at all. The first polarizer eliminates all but the horizontally polarized light waves. They are eliminated by the second polarizer, which passes only vertically polarized light. The result is that *no* light passes both the vertical and the horizontal polarizer. It does not matter whether the first polarizer is vertical and the second polarizer is horizontal or the other way round. The order of the filters is not important. In either case, no light passes through them.

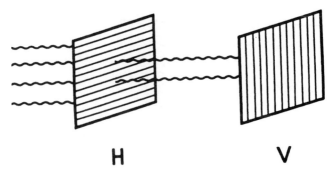

Whenever two polarizers are oriented at right angles to each other they block all light. No matter how the pair is twisted or turned as a unit, as long as they remain at right angles to each other, no light passes through them.

With this in mind, we come now to the third filter. Align the third filter so that it polarizes light *diagonally* and place it in front of the horizontal polarizer and the vertical polarizer. Nothing happens. If the first two filters (the horizontal and the vertical polarizer) block all the light, the addition of a third filter, of course, scarcely can affect the situation.

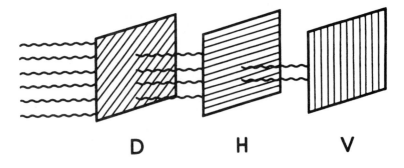

In a similar manner, if we place the diagonal polarizer on the other side of the horizontal-vertical combination, nothing happens. No light gets through the filters.

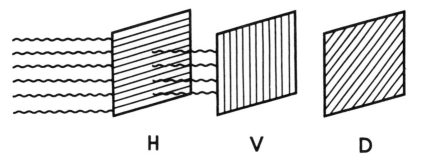

Now we come to the interesting part. Put the diagonal polarizer *between* the horizontal polarizer and the vertical polarizer. *Light gets through the three filters when cut off!*

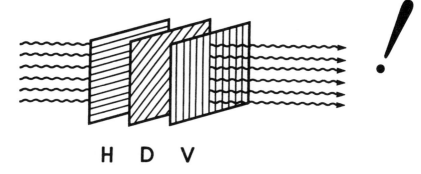

In other words, a combination of horizontal and vertical polarizers is as much a barrier to light waves as a wooden door. A diagonal polarizer in front of or behind such a combination does not affect this phenomenon. However, if a diagonal polarizer is sandwiched *in between* the horizontal polarizer and the vertical polarizer, light gets through all three of them. Remove the diagonal polarizer and the light disappears again. It is blocked by the combination of horizontal and vertical polarizers.

Diagrammatically, the situation looks like this:

How can this happen? According to quantum mechanics, diagonally polarized light is not a *mixture* of horizontally polarized light and vertically polarized light. We cannot simply say that the horizontal components of the diagonally polarized light passed through the horizontal polarizer and the vertical components of the diagonally polarized light passed through the vertical polarizer. According to quantum mechanics, diagonally polarized light is a separate thing-in-itself. How can a separate thing-in-itself get through all three filters but not through two of them?

If we consider light as a particle phenomenon, this paradox becomes graphic. Namely, how can *one photon* be broken down into a horizontally polarized component and a vertically polarized component? (By definition, it can't.)

This paradox is at the heart of the difference between quantum logic and classical logic. It is caused by our thought processes which are following the rules of classical logic. Our intellect tells us that what we are seeing is impossible (after all, a single photon must be polarized one way or the other). Nonetheless, whenever we insert a diagonal polarizer between a horizontal polarizer and a vertical polarizer, we see light where there was none before. Our eyes are ignorant of the fact that what they are seeing is "impossible." That is because experience does not follow the rules of classical logic. It follows the rules of quantum logic.

The thing-in-itselfness of diagonally polarized light reflects the true nature of *experience*. Our symbolic thought process imposes upon us the categories of "either-or." It confronts us always with either this or that, or a mixture of this and that. It says that polarized light is either vertically polarized or horizontally polarized or a mixture of vertically and horizontally polarized light. Thus are the rules of classical logic, the rules of *symbols*. In the realm of *experience,* nothing is either this or that. There is always at least one more alternative, and often an unlimited number of them.

Finkelstein put it this way in reference to quantum theory:

There are no waves in the game. The equation that the game obeys is a wave equation, but there are no waves running around. (This is one of the mountains of quantum mechanics.) There are no particles running around either. What's running around are quanta, the third alternative.[10]

To be less abstract, imagine that we have two different pieces from a chess set, say a bishop and a pawn. If these macroscopic chess pieces followed the same rules as quantum phenomena, we would not be able to say that there is nothing between being either a bishop or a pawn. Between the extremes of "bishop" and "pawn" is a creature called a "bishawn." A "bishawn" is neither a bishop nor a pawn nor is it half a bishop and half a pawn glued together. A "bishawn" is a separate thing-in-itself. It cannot be separated into its pawn component and its bishop component any more than a puppy which is half collie and half German shepherd can be separated into its collie "component" and its German shepherd "component."

There is more than one type of "bishawn" between the extremes of bishop and pawn. The bishawn that we have been describing is one-half part bishop and one-half part pawn. Another type of bishawn is one-third part bishop and two-thirds part pawn. Still another type of bishawn is three-fourths part bishop and one-fourth part pawn. In fact, for every possible proportion of parts bishop to parts pawn there exists a bishawn which is quite distinct from all the others.

A "bishawn" is what physicists call a coherent superposition. A "superposition" is one thing (or more) imposed on another. A double exposure, the bane of careless photographers, is a superposition of one photograph on another. A coherent superposition, however, is not simply the superposition of one thing on another. *A coherent superposition is a thing-in-itself which is as distinct from its components as its components are from each other.*

Diagonally polarized light is a coherent superposition of horizontally polarized light and vertically polarized light. Quantum physics abounds with coherent superpositions. In fact, coherent superpositions are at the heart of the mathematics of quantum mechanics. Wave functions are coherent superpositions.

Every quantum mechanical experiment has an observed system. Every observed system has an associated wave function. The wave function of a particular observed system (like a photon) is the coherent superposition of all the possible results of an interaction between the observed system and a measuring system (like a photographic plate). The development in time of this coherent superposition of possibilities is described by Schrödinger's wave equation. Using this equation, we can calculate the form of this thing-in-itself, this coherent superposition of possibilities which we call a wave function, for any given time. Having done *that,* we then can calculate the *probability* of each possibility contained in the wave function at that particular time. This gives us a probability function, which is not the same as a wave function, but is calculated from a wave function. In a nutshell, that is the mathematics of quantum physics.

In other words, in the mathematical formulations of quantum theory nothing is either "this" or "that" with nothing in between. Graduate students in physics routinely learn the mathematical technique of superimposing every "this" on every "that" in such a way that the result is neither the original "this" nor the original "that," but an entirely new thing called a coherent superposition of the two.

According to Finkelstein, one of the major conceptual difficulties of quantum mechanics is the false idea that these wave functions (coherent superpositions) are real *things* which develop, collapse, etc. On the other hand, the idea that coherent superpositions are pure abstractions which represent nothing that we encounter in our daily lives also is incorrect. They reflect the nature of *experience.*

How do coherent superpositions reflect experience? Pure experience is never restricted to merely two possibilities. Our *conceptualization* of a given situation may create the illusion that each dilemma has only two horns, but this illusion is caused by assuming that experience is bound by the same rules as symbols. In the world of symbols, everything is either this or that. In the world of experience there are more alternatives available.

For example, consider the judge who must try his own son in a court of law. The law allows only two verdicts: "He is guilty" and "He is innocent." For the judge, however, there is another possible verdict, namely, "He is my son." The fact that we prohibit judges from trying cases in which they have a personal interest is a tacit admission that experience is not limited to the categorical alternatives of "guilty" and "innocent" (or "good" and "bad," etc.). Only in the realm of symbols is the choice so clear.

During the Lebanese civil war, a story goes, a visiting American was stopped by a group of masked gunmen. One wrong word could cost him his life.

"Are you Christian or Moslem?" they asked.

"I am a *tourist!*" he cried.

The way that we pose our questions often illusorily limits our responses. In this case, the visitor's fear for his life broke through this illusion. Similarly, the way that we *think our thoughts* illusorily limits us to a perspective of either/or. Experience itself is never so limited. There is always an alternative between every "this" and every "that." The recognition of this quality of experience is an integral part of quantum logic.

Physicists engage in a particular kind of dance which is foreign to most of us. Being around them for any length of time is like entering another culture. Within this culture every statement is subject to the challenge, "Prove it!"

When we tell a friend, "I feel great this morning," we do not expect him to say, "Prove it." However, when a physicist says, "Experience is not bound by the same rules as symbols," he invites a chorus of "Prove it"s. Until he can do this, he must preface his remarks with, "It is my opinion that . . ." Physicists are not very impressed with opinions. Unfortunately, this sometimes makes them narrow-minded to an extraordinary degree. If you are not willing to follow *their* dance, they won't dance with you.

Their dance requires a "proof" for every assertion. A "proof" does not verify that an assertion is "true" (that that is the way the world really is). A scientific "proof" is a mathematical demonstration that the assertion in question is logically consistent. In the realm of pure mathematics, an assertion may have no relevance to experience at all. Nonetheless, if it is accompanied by a self-consistent "proof" it is accepted. If it is not, it is rejected. The same is true of physics except that the science of physics imposes the additional requirement that the assertion relate to physical reality.

So much for the relationship between the "truth" of a scientific assertion and the nature of reality. There isn't any. Scientific "truth" has nothing to do with "the way that reality really is." A scientific theory is "true" if it is self-consistent and correctly correlates experience (predicts events). In short, when a scientist says that a theory is true, he means that it correctly correlates experience and, therefore, it is useful. If we substitute the word "useful" whenever we encounter the word "true," physics appears in its proper perspective.

Birkhoff and von Neumann created a "proof" that experience violates the rules of classical logic. This proof, of course, is embedded in experience. In particular, it is based upon what does and does not happen with various combinations of polarized light. Finkelstein uses a slightly modified version of Birkhoff and von Neumann's original proof to demonstrate quantum logic.

The first step of this proof is to experiment with all of the possible combinations of horizontally, vertically, and diagonally polarized light. In other words, the first step is to do what we have done already: discover which light emissions pass through which polarizers. Observe for yourself that light passes through two vertical polarizers, two horizontal polarizers, two diagonal polarizers, a diagonal and a horizontal polarizer, and a diagonal and a vertical polarizer. All of these combinations are called "allowed transitions" because they actually happen. Similarly, observe for yourself that light does not pass through a horizontal and a vertical polarizer, or any other combination of polarizers oriented at right angles to each other. These combinations are called "forbidden transitions" because they never happen.

The second step of the proof is to make a table of this information called a transition table. A transition table looks like this:

ADMISSIONS

	)∅	)H	)V	)D	)D̄	)I
∅)						
H)		A		A	A	A
V)			A	A	A	A
D)		A	A	A		A
D̄)		A	A		A	A
I)		A	A	A	A	A

(left label: EMISSIONS)

The row of letters on the left are emissions. An emission is just what it sounds like. In this case, an emission is a light wave that is emitted from a light bulb. The ")" sign to the right of a letter indicates an emission. For example, "H)" means horizontally polarized light emitting from a horizontal polarizer. The row of letters on the top are admissions. An admission is the reception of an emission. The ")" sign to the left of a letter indicates an admission. For example, ")H" means a horizontally polarized light wave reaching an eyeball.

The zeros with the lines through them (pronounced "oh") stand for the "null process." The null process means that we decided to go to the movies today and not do the experiment. The null process stands for no emissions, nothing. The letter "I" stands for "identity process." The identity process is a filter that passes everything. In other words, "I" tells us what kinds of polarized light pass through, say, an open window: namely, every kind.

Two kinds of diagonally polarized light are included in the table to make it complete. The "D" represents light diagonally polarized to the left, and the "D̄" represents light diagonally polarized to the right (or the other way round).

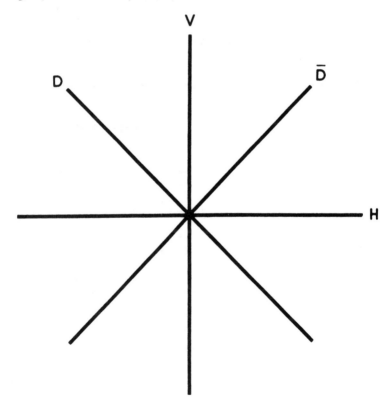

To use the transition table we pick the type of emission in which we are interested and follow it across the table. For example, an emission of horizontally polarized light, H), will pass through another horizontal polarizer, so an "A," for allowed, is placed in the square where the horizontally polarized emission line intersects the horizontally polarized admission column. Horizontally polarized light also passes through a diagonal polarizer tilted to the left,)D, a diagonal polarizer tilted to the right,)D̄, and an open window, I. An "A" is placed in each appropriate square.

Notice that the square where the horizontally polarized emission line intersects the vertically polarized admission column is blank. This is because horizontally polarized light does not pass through a vertical polarizer. The blank squares show the forbidden transitions. All of the null process squares are blank because nothing happens if we don't do the experiment. All of the "I" squares are marked "A" because every kind of light, polarized and otherwise, passes through an open window.

The third step in the proof is to make a simple diagram of the information contained in the transition table. The diagram made from this particular transition table looks like this:

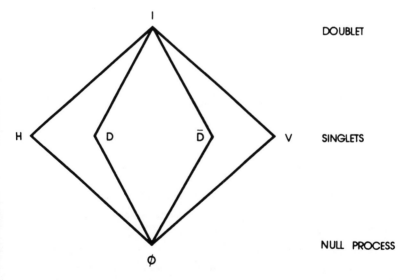

This type of diagram is called a lattice. Mathematicians use lattices to show the ordering of events or elements. Lattices are similar to the genealogical trees that we construct when we research our family roots. The higher elements include the lower elements. The lines show who is connected to whom and through whom.

A lattice is not exactly a family tree, but it shows the same kind of inclusive ordering. At the bottom is the null process. Nothing is below the null process since the null process represents no emissions of any kind. In the next level up are the various states of polarization. The elements at this level are called singlets. Singlets are the simplest statements that we can make about the polarization of a light wave. "This light is horizontally polarized," is the most that we can say about the state of polarization, even though it doesn't tell us anything else. It is a "maximal but incomplete description," a limitation inherent in the use of language.

The next level up contains the doublets. In this lattice there is only one doublet. Doublets comprise the next level of maximal but incomplete statements that we can make about the polarization of light in this simple experiment. Lattices representing more complex phenomena can have considerably more levels—triplets, quadruplets, etc. This lattice is the simplest of them all, but it graphically demonstrates the nature of quantum logic.

First, notice that the doublet, I, contains *four* singlets. This is typical of quantum logic but an incomprehensible contradiction to classical logic wherein every doublet, by definition, contains only *two* singlets, no more and no less. Lattices are graphic demonstrations of the quantum postulate that there is always at least one alternative between every "this" and every "that." In this case, two alternatives ("D" and "D̄") are represented. There are many more available alternatives that are not represented in this lattice. For example, the light in the lattice represented by the symbol D̄ is diagonally polarized at 45°, but we also can polarize light at 46°, 47°, $48^1/_2$°, etc., and all of these states of polarization could be included in the doublet, I.

In both classical logic and quantum logic a singlet can be represented by a dot. In classical logic a doublet is represented by two dots. In quantum logic, however, a doublet is represented by a line drawn between the two dots. *All of the points on the line* are included in the doublet—not only the two points that define it.

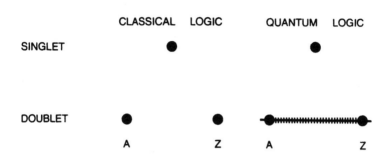

Now let us return to the distributive law: "A, and B or C" equals "A and B, or A and C." (The whole purpose of making a transition table was to make a lattice to use in disproving the distributive law.)

Mathematicians use lattice diagrams to determine which elements in the lattice are connected and in what way.

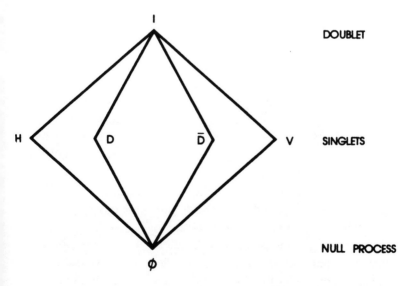

For example, to see how two elements in the lattice are connected by the word "and," follow the lines leading from the elements in question *down* to a point where they both meet (which mathematicians call the "greatest lower bound"). If we are interested in "H and D," we follow the lines *downward* from H and from D and find that they meet at Ø. Therefore, the lattice tells us that "H and D" equals "Ø." If we are interested in "I and H," we follow the line *downward* from the highest starting point on the lattice (I) and find that the lowest common point of I and H is H. Therefore, the lattice tells us that "I and H" equals "H."

To see how two elements in the lattice are connected by the word "or," follow the lines leading from the elements in question *up* to a point where they both meet (which mathematicians call the "least upper bound"). For example, if we are interested in "H or V," we follow the lines *upward* from H and from V and find that they meet at I. Therefore, the lattice tells us that "H or V" equals "I." Similarly, to find "D or I," we follow the lines *upward* to their highest common point, which is I. Therefore, the lattice tells us that "D or I" equals "I."

The rule is simple: "and" goes *down,* "or" goes *up.*

Go *down* the lattice to find "and," go *up* the lattice to find "or."

Now we come to the proof itself. The proof itself is considerably simpler than the preliminary explanations. The distributive law says that "A, and B or C" equals "A and B, or A and C." To see whether this is true of experience or not we simply insert some of our actual states of polarization into the formula and solve it using the lattice method. For example, the distributive law says that "Horizontally polarized light and vertically polarized light or diagonally polarized light" equals "Horizontally polarized light and vertically polarized light, or horizontally polarized light and diagonally polarized light." Using the abbreviations that we already have used, this is written: "H, and D or V" equals "H and D, or H and V."

Returning to the lattice, let us examine the left side of this statement first. Solving for "D or V," we follow the lines on the lattice upward from D and from V to their highest common point ("or" goes *up*). They meet at I. Therefore, the lattice tells us that "D or V" equals "I." Substituting "I" for the original "D or V," we have left

on this side of the statement "H and I." Following the lines from H
and from I downward on the lattice ("and" goes *down*), we find that
their lowest common point is at H. Therefore, the lattice tells us that
"H and I" equals "H."

In short:

"H, and D or V" equals "H and D, or H and V"
"H and I" equals "H and D, or H and V"
"H" equals "H and D, or H and V"

We solve the right side of this statement in the same way. Solving
for "H and D," we follow the lines on the lattice downward from H
and from D to their lowest common point. They meet at Ø. Therefore, the lattice tells us that "H and D" equals "Ø."

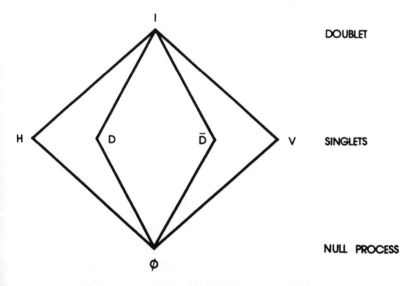

Substituting "Ø" for "H and D," we are left with "Ø, or H and
V" on the right side of the statement. To solve for "H and V," we
follow the lines on the lattice downward from H and from V to their
lowest common point. They also intersect at Ø. Therefore, the lattice
tells us that "H and V" equals "Ø." Substituting "Ø" for "H and V"

we now are left with "Ø or Ø" on the right side of the original statement. Both the lattice and common sense tell us that "Ø or Ø" equals
"Ø."

In short:

"H" equals "H and D, or H and V"
"H" equals "Ø, or H and V"
"H" equals "Ø or Ø"
"H" equals "Ø"

But "H" does *not* equal "Ø!" "H" is horizontally polarized
light and "Ø" is a non-experiment—a lack of any emission at all. The
distributive law does not work!

Here is Birkhoff and von Neumann's proof again. It is important
because, simple as it is, it could end an illusion millennia old: the
illusion that symbols and experience follow the same set of rules.
Except for the mathematical symbols that represent the connectives
"and" and "or," this is exactly the way that physicists read it:

$$\text{"H, and D or V"} \overset{?}{=} \text{"H and D, or H and V"}$$
$$\text{"H and I"} \overset{?}{=} \text{"Ø, or H and V"}$$
$$\text{"H"} \overset{?}{=} \text{"Ø or Ø"}$$
$$\text{"H"} \neq \text{"Ø"}$$

Finkelstein's theory is a theory of *process.* Quantum logic is only
one part of it. According to this theory, the basic unit of the universe
is an event, or a process. These events link in certain ways (allowed
transitions) to form webs. The webs in turn join to form larger webs.
Farther up the ladder of organization are coherent superpositions of
different webs (things which are neither "this web" nor "that web"
but distinct entities in themselves).

The basic events of Finkelstein's theory do not exist in space and
time. They are *prior* to space and time. According to Finkelstein,
space, time, mass, and energy are secondary qualities which are derived

from the basic events of the universe. In fact, Finkelstein's latest paper is called "Beneath Time."

This bold theory is a radical departure from conventional physics and from conventional thought. The mathematics of Finkelstein's theory, which is called quantum topology, is remarkably simple compared to the complex mathematics of quantum theory and relativity. Quantum topology is as yet incomplete (lacking "proof"). Like many theories, it may never be complete. Unlike most other theories, however, it contains the potential to alter radically our conceptual framework.

Von Neumann's discovery that our thought processes (the realm of symbols) project illusory restrictions onto the real word is essentially the same discovery that led Einstein to the general theory of relativity. Einstein disproved the universality of Euclidean geometry. Until the general theory of relativity, Euclidean geometry had been accepted without question as the underlying structure of the universe. Birkhoff and von Neumann disproved the universality of classical logic. Until now, classical logic has been accepted without question as a natural reflection of the nature of reality.

A powerful awareness lies dormant in these discoveries: an awareness of the hitherto-unsuspected powers of the mind to mold "reality," rather than the other way round. In this sense, the philosophy of physics is becoming indistinguishable from the philosophy of Buddhism, which is the philosophy of enlightenment.

1

The End of Science

A vital aspect of the enlightened state is the experience of an all-pervading unity. "This" and "that" no longer are separate entities. They are different *forms* of the same thing. Everything is a *manifestation*. It is not possible to answer the question, "Manifestation of *what?*" because the "what" is that which is beyond words, beyond concept, beyond form, beyond even space and time. Everything is a manifestation of that which is. That which is, is. Beyond these words lies the experience; the experience of that which is.

The forms through which that which is manifests itself are each and every one of them perfect. *We* are manifestations of that which is. *Everything* is a manifestation of that which is. Everything and everybody is exactly and perfectly what it is.

A fourteenth-century Tibetan Buddhist, Longchenpa, wrote:

> Since everything is but an apparition
> Perfect in being what it is,
> Having nothing to do with good or bad,
> Acceptance or rejection,
> One may well burst out in laughter.[1]

We might say, "God's in His heaven and all's well with the world," except that according to the enlightened view, the world couldn't be any other way. It is neither well nor not well. It simply is what it is. What it is is perfectly what it is. It couldn't be anything else. It is perfect. I am perfect. I am exactly and perfectly who I am. You are perfect. You are exactly and perfectly who you are.

If you are a happy person, then that is what you perfectly are— a happy person. If you are an unhappy person, then *that* is what you perfect are—an unhappy person. If you are a person who is changing, then *that* is what you perfectly are—a person who is changing. That which is is that which is. That which is not is that which is. There is nothing which is not that which is. There is nothing other than that which is. Everything is that which is. We are a part of that which is. In fact, *we are that which is.*

If we substitute "subatomic particles" for people in this scheme, we have a good approximation of the conceptual dynamics of particle physics. Yet, there is another sense in which this aspect of unity has entered physics. The pioneers of quantum physics noticed a strange "connectedness" among quantum phenomena. Until recently this oddity lacked any theoretical significance. It was regarded as an accidental feature which would be explained as the theory developed.

In 1964, J. S. Bell, a physicist at the European Organization for Nuclear Research (CERN) in Switzerland, zeroed in on this strange connectedness in a manner that may make it the central focus of physics in the future. Dr. Bell published a mathematical proof which came to be known as Bell's theorem. Bell's theorem was reworked and refined over the following ten years until it emerged in its present form. Its present form is dramatic, to say the least.

Bell's theorem is a mathematical construct which, as such, is indecipherable to the nonmathematician. Its implications, however, could affect profoundly our basic world view. Some physicists are convinced that it is the most important single work, perhaps, in the history of physics. One of the implications of Bell's theorem is that, at a deep and fundamental level, the "separate parts" of the universe are connected in an intimate and immediate way.

In short, Bell's theorem and the enlightened experience of unity are very compatible.

The unexplained connectedness of quantum phenomena shows itself in several ways. The first way we already have discussed. It is the double-slit experiment (page 66). When both slits in a double-slit experiment are open, the light waves going through them interfere with each other to form a pattern of alternating light and dark bands on a screen. When only one slit in a double-slit experiment is open, the light waves going through it illuminate the screen in the ordinary way. How does a single photon in a double-slit experiment know whether or not it can go to an area on the screen that must be dark if both slits are open?

The great multitude of photons of which a single photon eventually will be a part distributes itself in one way if one slit is open, and in an entirely different way if both slits are open. The question is, assuming that a single photon goes through one of the two slits, *how does it know whether or not the other slit is open?* Somehow it does. An interference pattern *always* forms when we open both slits, and it *never* forms when we close one of the slits.

However, there is another experiment in which this apparent connectedness of quantum phenomena is even more perplexing. Suppose that we have what physicists call a two-particle system of zero spin. This means that the spin of each of the particles in the system cancels the other. If one of the particles in such a system has a spin *up*, the other particle has a spin *down*. If the first particle has a spin *right*, the second particle has a spin *left*. No matter how the particles are oriented, their spins are always equal and opposite.

Now suppose that we separate these two particles in some way that does not affect their spin (like electrically). One particle goes off in one direction and the other particle goes off in the opposite direction.

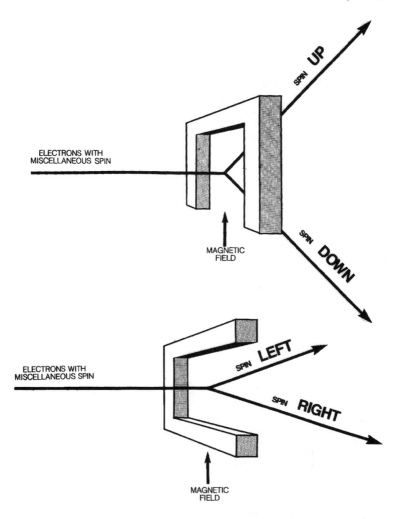

The spin of a subatomic particle can be oriented by a magnetic field. For example, if a beam of electrons with randomly oriented spin is sent through a particular type of magnetic field (called a Stern-Gerlach device), the magnetic field splits the beam into two equal smaller beams. In one of them all of the electrons have a spin *up* and in the other all of the electrons have a spin *down*. If only one electron goes through this magnetic field, it will come out of it with either a spin *up* or a spin *down*. (We can design the experiment so that the odds are 50-50) (first drawing, previous page).

If we reorientate the magnetic field (change its axis) we can give all of the electrons a spin *right* or a spin *left* instead of a spin up or a spin down. If only one electron goes through the magnetic field when it is oriented this way, it will come out of it with either a spin *right* or a spin *left* (equal chance either way) (second drawing, previous page).

Now suppose that after we separate our original two-particle system we send one of the particles through a magnetic field that will give it either a spin *up* or a spin *down*. In this case, let us say that the particle comes out of the magnetic field with a spin *up*. This means that we automatically know that the other particle has a spin *down*. We do not have to make a measurement on the other particle because we know that its spin is equal to and opposite to that of its twin.

The experiment looks like this:

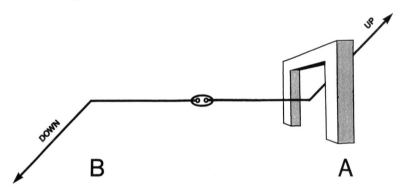

The original two-particle system with zero spin is at the center. One of the particles goes to area A. In area A it goes through a Stern-Gerlach device. In this case, the Stern-Gerlach device gives it a spin *up*. Therefore, we know *without measuring* that the other particle, which has gone to area B, has a spin *down*.

Albert Einstein, Boris Podolsky, and Nathan Rosen thought up this experience over forty-five years ago. Actually, this version of the Einstein-Podolsky-Rosen experiment (using spin states) was thought up by David Bohm, a physicist at the University of London. This version usually is used to illustrate the Einstein-Podolsky-Rosen effect. (The original paper dealt with positions and momenta.)

In 1935, Einstein, Podolsky, and Rosen published their thought experiment in a paper entitled, "Can Quantum-Mechanical Description of Physical Reality Be Considered Complete?"[2] At that time, Bohr, Heisenberg, and the proponents of the Copenhagen Interpretation of Quantum Mechanics (page 40) were saying that quantum theory is a complete theory even though it doesn't provide any picture of the world separate from our observations of it. (They're still saying it.) The message that Einstein, Podolsky, and Rosen wanted to convey to their colleagues was that the quantum theory is *not* a "complete" theory because it does not describe certain important aspects of reality which are physically real even if they are not observed. The message that their colleagues got, however, was quite different. The message that their colleagues got was that the particles in the Einstein-Podolsky-Rosen thought experiment somehow are connected in a way that transcends our usual ideas about causality.

For example, if the axis of the Stern-Gerlach device in our hypothetical experiment were changed to make the particles spin *right* or *left* instead of *up* or *down,* the experiment would look like this:

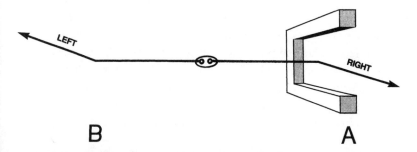

The particle in area A would have a spin *right* instead of a spin *up.* This means that the particle in area B would have a spin *left* instead of a spin *down.* Its spin is always equal and opposite that of its twin.

Now suppose that the axis of the Stern-Gerlach device is changed while the particles are *in flight.* Somehow the particle traveling in area B "knows" that its twin in area A is spinning *right* instead of *up* and so it spins *left* instead of *down*. In other words, *what we did in area A (changed the axis of the magnetic field) affected what happened in area B.* This strange phenomenon is known as the Einstein-Podolsky-Rosen (EPR) effect.

Einstein, Podolsky, and Rosen's thought experiment is the Pandora's Box of modern physics. It inadvertently illustrated an unexplainable connectedness between particles in two different places. The particle in area B seems to know *instantaneously* the spin status of the particle in area A.* This connectedness allows an experimenter in one place (area A) to affect the state of a system in another place (area B).

"It is rather discomforting," remarked Erwin Schrödinger, in reference to this phenomenon,

> that the [quantum] theory should allow a system to be steered or piloted into one or the other type of state at the experimenter's mercy in spite of his having no access to it.[3]

At once physicists realized that this peculiar situation raises a critical question: "How can two of anything communicate so quickly?"

According to the usual ideas in physics, information is carried from one place to another by a signal. Without a carrier there is no communication. For example, the most common form of communication is talking. The information that we convey by talking is carried (in a face-to-face conversation) by sound waves. Sound waves only travel so fast (about 700 miles per hour). Therefore, how long it takes my information to get from me to you depends upon how far away from me you are. The fastest communication signal is an electromagnetic wave, like a light wave or a radio wave. These travel at approxi-

---

*As seen from a particular co-ordinate system. We must be careful about using words like "instantaneous." Einstein's special theory of relativity shows that whether one event appears to occur simultaneously with, before, or after another event depends upon the frame of reference from which the observation is made. Accurately speaking, this type of communication is called "space-like" (see next page). Space-like transfers do not always appear to be instantaneous. In fact, they only appear instantaneous from special frames of reference.

mately 186,000 miles per second. Almost all of physics rests upon the assumption that *nothing in the universe can travel faster than the speed of light.** The extraordinarily high velocity of light makes communication by light signal *seem* instantaneous. I seem to see you nod your head at the moment that you actually do it. Nonetheless, communication by light signal is not instantaneous. How long it takes my information to travel via light signal from me to you depends upon how far away from me you are. In most instances, the travel time required is so brief that it scarcely can be measured. It takes several seconds, however, for a radio signal to travel from the earth to the moon and back.

Now suppose that area A and area B are very far apart. It will take a certain amount of time for a light signal to travel from area A to area B. If area A and area B are so far apart that there is insufficient time for a light signal to connect an event that happens in area A with an event that happens in area B, there is no way, according to the usual ideas in physics, that the event in area B can know about the event in area A. Physicists call this a "space-like" separation. (One event is space-like separated from another event if there is insufficient time for a light signal to connect them.) Communication between space-like separated events defies one of the most basic assumptions of physics. Yet this is exactly what the Einstein-Podolsky-Rosen thought experiment seems to illustrate. Even though they are space-like separated, the state of the particle in area B depends upon what the observer in area A decides to observe (which way he orients his magnetic field).

In other words, the Einstein-Podolsky-Rosen effect indicates that information can be communicated at *superluminal* (faster than light) speeds contrary to the accepted ideas of physicists. If the two particles in the Einstein-Podolsky-Rosen thought experiment some-

*Relativity permits the hypothetical existence of particles called tachyons (tak'i ons) which come into existence already traveling faster than light. In the formalism of the special theory of relativity, tachyons have an imaginary rest mass. Unfortunately, no one knows what an "imaginary rest mass" means in physical terms, or what the interaction forces would be between tachyons and the ordinary particles of real rest mass out of which we are made.

how are connected by a signal, that signal is traveling faster than the speed of light. Einstein, Podolsky, and Rosen may have created the first scientific example of a superluminal connection.

Einstein himself denied this conclusion. It is not possible, he argued, that the setting we choose for a measuring device here can affect what happens somewhere else. In his autobiography, written eleven years after the Einstein-Podolsky-Rosen paper, he wrote:

> . . . on one supposition we should, in my opinion, absolutely hold fast; the real factual situation of the system S_2 [the particle in area B] is independent of what is done with the system S_1 [the particle in area A], which is spatially separated from the former.[4]

This opinion is, in effect, the principle of local causes. The principle of local causes says that what happens in one area does not depend upon variables subject to the control of an experimenter in a distant space-like separated area. The principle of local causes is common sense. The results of an experiment in a place distant and space-like separated from us should not depend on what we decide to do or not to do right here. (Except for the mother who rose in alarm at the same instant that her daughter's distant automobile crashed into a tree—and similar cases—the macroscopic world appears to be made of local phenomena.)

Since phenomena are local in nature, argued Einstein, quantum theory has a serious flaw. According to quantum theory, changing the measuring device in area A changes the wave function which describes the particle in area B, but (according to Einstein) it cannot change "the real factual situation of the system S_2 [which] is independent of what is done with the system S_1 . . ."

Therefore, one and the same "factual situation" in area B has two wave functions, one for each position of the measuring device in area A. This is a flaw since it is "impossible that two different types of wave functions could be coordinated with the identical factual situation of S_2."[5]

Here is another way of looking at the same situation: Since the real factual situation in area B is independent of what is done in area A, there must exist *simultaneously* in area B a definite spin *up* or *down* and a definite spin *right* or *left* to account for all the results that we can get by orienting the Stern-Gerlach device in area A either vertically or horizontally. Quantum theory is not able to describe such a state in area B and, therefore, it is an incomplete theory.*

However, Einstein closed his argument with an incredible aside:

> One can escape from this conclusion [that quantum theory is incomplete] only by either assuming that the measurement of S_1 ((telepathically)) changes the real situation of S_2 or by denying independent real situations as such to things which are spatially separated from each other. Both alternatives appear to me entirely unacceptable.[6]

Although these alternatives were unacceptable to Einstein, they are being considered by physicists today. Few physicists believe in telepathy, but some physicists do believe either that at a deep and fundamental level there is no such thing as "independent real situations" of things which have interacted in the past but which are spa-

---

*The EPR argument for the incompleteness of quantum theory rests squarely on the assumption that the real factual situation in one region cannot depend on what an experimenter does in a far-away region (the principle of local causes).

Einstein, Podolsky, and Rosen point out that we could have chosen to place the axis of the magnet in area A either in the vertical position or in the horizontal position, and that in each case we would have observed a definite result—either *up* or *down* in the vertical case, or *right* or *left* in the horizontal case. They also assert that what we do (choose to measure or observe) in area A cannot affect the real factual situation in area B. Thus they conclude there must exist *simultaneously* in area B a definite spin, *up* or *down,* and also a definite spin, *right* or *left,* to account for all the possible results that we can get by orienting the magnet in area A one way or the other.

Quantum theory is not able to describe such a state and hence Einstein, Podolsky, and Rosen concluded that the description which quantum theory provides is not complete; the quantum description cannot represent certain information about the system in area B (the simultaneous existence of different spin states) that is needed to completely describe the situation there.

tially separated from each other, or that changing the measuring device in area A *does* change "the real factual situation" in area B.

This brings us to Bell's theorem.

Bell's theorem is a mathematical proof. What it "proves" is that if the statistical predictions of quantum theory are correct, then some of our commonsense ideas about the world are profoundly mistaken.

Bell's theorem does not demonstrate clearly in what way our commonsense ideas about the world are inadequate. There are several possibilities. Each possibility has champions among the small number of physicists who are familiar with Bell's theorem. No matter which of the implications of Bell's theorem we favor, however, Bell's theorem itself leads to the inescapable conclusion that if the statistical predictions of quantum theory are correct, then our commonsense ideas about the world are profoundly deficient.

This is quite a conclusion because *the statistical predictions of quantum mechanics are always correct.* Quantum mechanics is *the* theory. It has explained everything from subatomic particles to transistors to stellar energy. It never has failed. It has no competition.

Quantum physicists realized in the 1920s that our commonsense ideas were inadequate for describing subatomic phenomena (pages 20, 260). Bell's theorem shows that commonsense ideas are inadequate even to describe macroscopic events, events of the everyday world!

As Henry Stapp wrote:

> The important thing about Bell's theorem is that it puts the dilemma posed by quantum phenomena clearly into the realm of macroscopic phenomena . . . [it] shows that our ordinary ideas about the world are somehow profoundly deficient even on the macroscopic level.[7]

Bell's theorem has been reformulated in several ways since Bell published the original version in 1964. No matter how it is formulated, it projects the "irrational" aspects of subatomic phenomena squarely into the macroscopic domain. It says that not only do events in the realm of the very small behave in ways which are utterly different from our commonsense view of the world, but also that events in the world at large, the world of freeways and sports cars, behave in ways which are utterly different from our commonsense view of them. This incredible statement cannot be dismissed as fantasy because it is based upon the awesome and proven accuracy of the quantum theory itself.

Bell's theorem is based upon correlations between paired particles similar to the pair of hypothetical particles in the Einstein-Podolsky-Rosen thought experiment.* For example, imagine a gas that emits light when it is electrically excited (think of a neon sign). The excited atoms in the gas emit photons in pairs. The photons in each pair fly off in opposite directions. Except for the difference in their direction of travel, the photons in each pair are identical twins. If one of them is polarized vertically, the other one also is polarized vertically. If one of the photons in the pair is polarized horizontally, the other photon also is polarized horizontally. No matter what the angle of polarization, both photons in every pair are polarized in the same plane.

Therefore, if we know the state of polarization of one of the particles, we automatically know the state of polarization of the other particle. This situation is similar to the situation in the Einstein-Podolsky-Rosen thought experiment, except that now we are discussing states of polarization instead of spin states.

We can verify that both photons in each pair of photons are polarized in the same plane by actually sending them through polarizers. On the next page is a picture of this (conceptually) simple procedure.

---

*The original version of Bell's theorem involves spin $1/2$ particles. Clauser and Freedman's experiment (keep reading), like this one, involves photons.

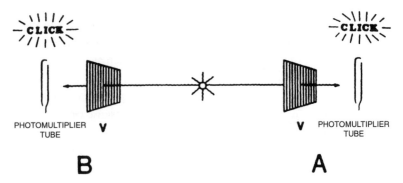

A light source in the center of the picture emits a pair of photons. On each side of the light source a polarizer is placed in the path of the emitted photon. Behind the polarizers are photomultiplier tubes which emit a click (or an audible electronic equivalent) whenever they detect a photon.

Whenever the photomultiplier tube in area A emits a click, the photomultiplier tube in area B also emits a click. This is because both of the photons in each photon pair always are polarized in the same plane, and both of the polarizers in this arrangement are aligned in the same direction (in this case, vertically). There is no theory involved here, just a matter of counting clicks. We know, and can verify, that when the polarizers both are aligned in the same direction, the photo-multiplier tubes behind them will click an equal number of times. The clicks in area A are correlated with the clicks in area B. The correlation, in this case, is *one*. Whenever one of the photomultiplier tube clicks, the other photomultiplier tube always clicks as well.

Now suppose that we orient one of the polarizers at 90 degrees to the other. On the next page is a picture of this arrangement. One of the polarizers still is aligned vertically, but the other polarizer now is aligned horizontally. Light waves that pass through a vertical polar-izer are stopped by a horizontal polarizer and the other way round. Therefore, when the polarizers are oriented at right angles to each other, a click in area A *never* will be accompanied by a click in area B. The clicks in area A, again, are correlated with the clicks in area B. This time, however, the correlation is *zero*. Whenever one of the photo-multiplier tubes clicks, the other photomultiplier tube *never* clicks.

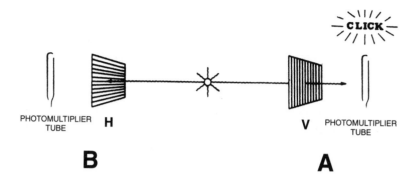

There also are correlations between the clicks in area A and the clicks in area B for every other possible combination of polarizer settings between these two extremes. These statistical correlations can be predicted by the quantum theory. For a given setting of the polarizers, a certain number of clicks in one area will be accompanied by a certain number of clicks in the other area.

Bell discovered that no matter what the settings of the polarizers, the clicks in area A are correlated too strongly to the number of clicks in area B to be explained by chance. They have to be connected somehow. However, if they are connected, then the principle of local causes (which says that what happens in one area does not depend upon variables subject to the control of an experimenter in a distant space-like area) is an illusion! In short, Bell's theorem shows that the principle of local causes, however reasonable it sounds, is mathematically incompatible with the assumption that the statistical predictions of quantum theory are valid (at least valid in this experiment and in the Einstein-Podolsky-Rosen experiment).*

---

*The Einstein-Podolsky-Rosen argument for the incompleteness of quantum theory was based upon the assumption of local causes. This assumption seemed plausible to most physicists because most physicists doubted that the real factual situation in one area of the EPR experiment actually was being influenced by the actions of a faraway observer. Their doubts arose from the fact that the quantum state composed of equal parts *up* and *down* is exactly equivalent to the quantum state composed of equal parts *right* and *left*. These two combinations are experimentally indistinguishable. Thus the actions of the far-away observer, by

The correlations which Bell used were calculated, but untested predictions of the quantum theory. In 1964, this experiment was still a hypothetical construct. In 1972, John Clauser and Stuart Freedman at the Lawrence Berkeley Laboratory actually performed this experiment to confirm or disprove these predictions.[8] They found that the statistical predictions upon which Bell based his theorem *are* correct.

Bell's theorem not only suggests that the world is quite different than it seems, it *demands* it. There is no question about it. Something very exciting is happening. Physicists have "proved," rationally, that our rational ideas about the world in which we live are profoundly deficient.

In 1975, Henry Stapp, in a work supported by the U.S. Energy Research and Development Administration, wrote:

> Bell's theorem is the most profound discovery of science.[9]

The deduction of superluminal communication from the results of the Clauser-Freedman experiment rests upon an important assumption: namely, that the states of the measuring devices prior to the arrival of the photons in area A and area B do not matter. This is, after all, a reasonable assumption. Normally we say that the orientation of a measuring device prior to a measurement is not relevant to the result

---

themselves, can have no observable effects here. Hence it is not clear that the real factual situation here is being changed.

The Einstein-Podolsky-Rosen argument (and the principle of local causes) was demolished by Bell in 1964. Bell showed that several assumptions that are implicit in the Einstein-Podolsky-Rosen argument imply that what happens experimentally in area B *must depend* on what the experimenter does in area A, or vice versa. The sufficient assumptions are: (1) that the observer in each area can orient the magnetic field in his area in either of the two alternative directions; (2) that some particular (although generally unknown) experimental result can be assumed to occur in each of the four alternative experimental situations; and (3) that the statistical predictions of quantum theory are valid (say to within 3%) in each of the four alternative cases. Bell's argument demonstrates by simple arithmetic that these three assumptions imply that the experimental results in one of the two areas must depend on what the observer in the other area chooses to observe (i.e., on how he orients the magnetic field of his Stern-Gerlach device). This conclusion contradicts the locality assumption of the Einstein-Podolsky-Rosen argument.

that we get at the time of a measurement. The result of an experiment depends upon the state of the measuring device at the time that the particle is detected by it, and not on the state of the measuring device before the particle gets there. However, superluminal communication cannot be deduced from the results of the Clauser-Freedman experiment without this assumption. Even though the photons in the photon pair cannot exchange information via light signals while they are in flight (each is traveling away from the other at the speed of light), the measuring device in area A and the measuring device in area B, which are set prior to the beginning of the experiment, may have exchanged information in the conventional manner (via light signals propagating within space-time). In other words, in the Clauser-Freedman experiment the information about the setting of the measuring device in either region has sufficient time, traveling at the speed of light or less, to reach the other region before the particle arrives.

In 1982, Alain Aspect, a physicist at the Institute of Optics, University of Paris, in Orsay, France, conducted an experiment which was similar to the Clauser-Freedman experiment, but with one important difference: the settings on the measuring devices in Aspect's experiment could be changed at the last minute (or, more precisely, at the last microsecond).[10] Changing the settings on the measuring devices at the last minute insures that information about the setting of the measuring device in either area does not have sufficient time, traveling at the speed of light or less, to reach the other region before the particle arrives.* In other words, Aspect, in effect, performed Bohm's thought experiment.

Like the Clauser-Freedman experiment (and several Clauser-Freedman-type experiments which had been performed in the meanwhile),[11] Aspect's experiment verified the statistical predictions of quantum mechanics. Because Aspect's experiment, however, satisfied the conditions upon which the logical analysis leading to the phenom-

---

*A tacit assumption is made here that the choice of the orientation of the polarization detector, which can be made by a random quantum process (like radioactive decay), or by the free will of an experimenter, has no deterministic roots in the past which are important in this context.

enon of superluminal transfer of information is based (that area A and area B are space-like separated) physicists were able to deduce this phenomenon solely on the basis of Aspect's experimental results. This lent considerable credence to the conclusion which Stapp had reached five years previously. Wrote Stapp:

> Quantum phenomena provide *prima facie* evidence that information gets around in ways that do not conform to classical ideas. Thus, the idea that information is transferred superluminally is, *a priori*, not unreasonable.
>
> Everything we know about Nature is in accord with the idea that the fundamental process of Nature lies outside space-time . . . but generates events that can be located in space-time. The theorem of this paper supports this view of Nature by showing that superluminal transfer of information is necessary, barring certain alternatives . . . that seem less reasonable. Indeed, the reasonable philosophical position of Bohr seems to lead to the rejection of the other possibilities, and hence, by inference, to the conclusion that superluminal transfer of information is necessary.[12]

Thus, eighty-two years after Planck presented his quantum hypothesis, physicists have been forced to consider the possibility, among others, that the superluminal transfer of information between space-like separated events may be an integral aspect of our physical reality.*,†

* * *

*"Superluminal transfer of information" refers to a fundamental phenomenon of nature, but not one that can be controlled, i.e., used to send messages. The possibility of utilizing superluminal information transfer to transmit encoded messages was proposed in 1975 by J. Sarfatti. However, H. Stapp and N. Herbert independently have pointed out, following Heisenberg (1929) and Bohm (1951), that any phenomenon which is described adequately by the quantum theory cannot be used to transmit messages superluminally.

†The phenomenon of superluminal information transfer between space-like separated events may be related to Jung's concept of synchronicity.

Bell's theorem showed that either the statistical predictions of quantum theory or the principle of local causes is false. It did not say which one is false, but only that both of them cannot be true. When Clauser and Freedman confirmed that the statistical predictions of quantum theory are correct, the startling conclusion was inescapable: The principle of local causes must be false! However, if the principle of local causes fails and, hence, the world is not the way it appears to be, then what is the true nature of our world?

There are several mutually exclusive possibilities. The first possibility, which we have just discussed, is that, appearances to the contrary, there really may be no such thing as "separate parts" in our world (in the dialect of physics, "locality fails"). In that case, the idea that events are autonomous happenings is an illusion. This would be the case for any "separate parts" that have interacted with each other at any time in the past. When "separate parts" interact with each other, they (their wave functions) become correlated (through the exchange of conventional signals) (forces). Unless this correlation is disrupted by other external forces, the wave functions representing these "separate parts" remain correlated forever.* For such correlated "separate parts," what an experimenter does in this area has an intrinsic effect upon the results of an experiment in a distant, space-like separated area. This possibility entails a faster-than-light communication of a type different than conventional physics can explain.

In this picture, what happens here is intimately and immediately connected to what happens elsewhere in the universe, which, in turn, is intimately and immediately connected to what happens elsewhere in the universe, and so on, simply because the "separate parts" of the universe are not separate parts.

"Parts," wrote David Bohm:

> are seen to be in immediate connection, in which their dynamical relationships depend, in an irreducible way, on the state of the whole system (and, indeed, on that of broader systems in

---

*If the Big Bang theory is correct, the entire universe is initially correlated.

which they are contained, extending ultimately and in principle to the entire universe). Thus, one is led to a new notion of *unbroken wholeness* which denies the classical idea of analyzability of the world into separately and independently existent parts . . .[13]

According to quantum mechanics, individual events are determined by pure chance (page 75). We can calculate, for example, that a certain percentage of spontaneous positive kaon decays will produce an antimuon and a neutrino (63%), a certain percentage will produce a positive pion and a neutral pion (21%), a certain percentage will produce two positive pions and a negative pion (5.5%), a certain percentage will produce a positron, a neutrino, and a neutral pion (4.8%), a certain percentage will produce an antimuon, a neutrino, and a neutral pion (3.4%), and so on. However, quantum theory cannot predict *which* decay will produce which result. Individual events, according to quantum mechanics, are completely random.

Said another way, the wave function which describes spontaneous kaon decays contains all of these possible results. When the decay actually happens, one of these potentialities is converted into an actuality. Even though the probability of each potentiality can be calculated, which potentiality actually happens at the moment of decay is a matter of chance.

Bell's theorem implies that which decay reaction occurs at a certain time is *not* a matter of chance. Like everything else, it is dependent upon something which is happening elsewhere.*

---

*The nonlocal aspect of nature illuminated by Bell's theorem is accommodated in quantum theory by the so-called collapse of the wave function. This collapse of the wave function is a sudden global change of the wave function of a system. It takes place when any part of the system is observed. That is, when an observation on the system is made in one region the wave function changes instantly, not only in that region, but also in far-away regions. This behavior is completely natural for a function that describes probabilities for probabilities depend on what is known about the system, and if knowledge changes as a result of an observation, then the probability function (the amplitude of the wave function squared) should change. Thus a change in the probability function in a distant

In the words of Stapp:

> . . . the conversion of potentialities into actualities cannot proceed on the basis of locally available information. If one accepts the usual ideas about how information propagates through space and time, then Bell's theorem shows that the macroscopic responses cannot be independent of far-away causes. This problem is neither resolved nor alleviated by saying that the response is determined by "pure chance." Bell's theorem proves precisely that the determination of the macroscopic response must be "nonchance," at least to the extent of allowing some sort of dependence of this response upon the far-away cause.[14]

Superluminal quantum connectedness seems to be, on the surface at least, a possible explanation for some types of psychic phenomena. Telepathy, for example, often appears to happen instantaneously, if not faster. Psychic phenomena have been held in disdain by physicists since the days of Newton. In fact, most physicists do not even believe that they exist.*

In this sense, Bell's theorem could be the Trojan horse in the physicists' camp; first, because it proves that quantum theory requires connections that appear to resemble telepathic communication, and second, because it provides the mathematical framework through which serious physicists (all physicists are serious) could find themselves discussing types of phenomena which, ironically, they do not believe exist.

The failure of the principle of local causes does not necessarily

---

region is normal even in classical physics. It reflects the fact that the parts of the system are correlated with each other, and hence, that an increase of information here is accompanied by an increase of information about the system elsewhere. However, in quantum theory this collapse of the wave function is such that what *happens* in a far-away place must, in some cases, depend on what an observer here chooses to observe, what you see there depends on what I *do* here. This is a completely nonclassical nonlocal effect.

*There are some notable exceptions, chief among which are Harold Puthoff and Russell Targ, whose experiments in remote viewing at the Stanford Research Institute are presented in their book, *Mind-Reach*, New York, Delacorte, 1977.

mean that superluminal connections actually exist. There are other ways to explain the failure of the principle of local causes. For example, the principle of local causes—that what happens in one area does not depend upon variables subject to the control of an experimenter in a distant space-like separated area—is based upon two tacit assumptions which are so obvious that they are easy to overlook.

First, the principle of local causes assumes that we have a choice about how we perform our experiments. Imagine that we are doing Clauser and Freedman's photon experiment. We have before us a switch which determines how the polarizers will be positioned. If we throw the switch up, the polarizers are aligned with each other. If we throw the switch down, the polarizers are oriented at right angles with each other. Suppose that we decide to throw the switch up and align the polarizers. Normally, we assume that *we could have* thrown the switch down and oriented the polarizers at right angles if we had wanted to. In other words, we assume that we are free to decide whether the switch before us will be up or down when the experiment begins.

The principle of local causes assumes (". . . variables subject to the control of an experimenter . . .") that we possess and can exercise a free will in the determination of how to perform our experiment. Second, and this is even easier to overlook, the principle of local causes assumes that if we had performed our experiment in a different way than we actually did perform it, we would have obtained some definite results. These two assumptions—that we can choose how to perform our experiment and that each of our choices, including those that we did not select, produces or would have produced definite results—is what Stapp calls "contrafactual definiteness."

The fact, in this case, is that we decided to perform our experiment with the switch in the "up" position. We assume that, contrary to this fact (contrafactually), we *could have* performed it with the switch in the "down" position. By performing the experiment with the switch in the "up" position, we obtained some definite results (a certain number of clicks in each area). Therefore, we assume that if we had chosen to perform the experiment with the switch in the "down" position, we likewise would have obtained some definite results. (It is not necessary that we be able to calculate what these other results are.) Odd as it may seem, some physical theories, as we shall see, do not assume that "what would have happened if . . ." produces definite results.

Since Bell's theorem shows that, assuming the validity of quantum theory, the principle of local causes is incorrect, and, if we do not want to accept the existence of superluminal connections ("the failure of locality") as the reason for the failure of the principle of local causes, then we are forced to confront the possibility that our assumptions about contrafactual definiteness are incorrect ("contrafactual definiteness fails"). Since contrafactual definiteness has two parts, there are two ways in which contrafactual definiteness could fail.

The first possibility is that free will is an illusion ("contrafactualness fails"). Perhaps there is no such thing as "what would have happened if . . ." Perhaps there can be only what is. In this case, we are led to a *superdeterminism*. This is a determinism far beyond ordinary determinism. Ordinary determinism states that once the initial situation of a system is established, the future of the system also is established since it must develop according to inexplorable laws of cause and effect. This type of determinism was the basis of the Great Machine view of the universe (page 24). According to this view, however, if the initial situation of a system, like the universe, is changed, then the future of the system also is changed.

According to superdeterminism, *not even the initial situation of the universe could be changed*. Not only is it impossible for things to be other than they are, it is even impossible that the initial situation of the universe could have been other than what it was. No matter what we are doing at any given moment, it is the only thing that *ever* was possible for us to be doing at that moment.

Contrafactual definiteness also fails if the "definiteness" assumption in it fails. In this case, we do have a choice in the way that we perform our experiments, but "what would have happened if . . ." does not produce any definite results. This alternative is just as strange as it sounds. It is also just what comes out of the Many Worlds Interpretation of Quantum Mechanics (page 92). According to the Many Worlds theory, whenever a choice is made in the universe between one possible event and another, the universe splits into different branches.

In our hypothetical experiment we decided to throw the switch into the "up" position. When the experiment was performed with the switch in the "up" position, it gave us a definite result (a certain number of clicks in each area). However, according to the Many Worlds theory, at the moment that we threw the switch up, the universe split into two branches. In one branch, the experiment was performed with the switch in the "up" position. In the other branch, the experiment was performed with the switch in the "down" position.

Who performed the experiment in the second branch? There is a different edition of *us* in each of the different branches of the universe! Each edition of us is convinced that *our* branch of the universe is the entirety of reality.

The experiment in the second branch, the experiment which was performed with the switch in the "down" position, also produced a definite result (a certain number of clicks in each area). However, that result is in another branch of the universe, not in ours. Therefore, as far as we in this branch of the universe are concerned, "what would have happened if . . ." actually *did* happen, and actually did produce definite results, but in a branch of the universe which is forever beyond our experiential reality.*

On the next page is a diagram of the logical implications of Bell's

---

*A branching also occurs at the choice between results. This can be illustrated by the EPR experiment. On the original branch where, for example, the axis of the magnetic field is vertical and the result is either spin *up* or spin *down*, there is a branching into two "twigs." In the first twig the result is spin *up*, and in the second twig the result is spin *down*. Similarly, on the second branch, where the axis of the magnetic field is horizontal, there is also a branching into two twigs. On the first of these twigs the result is spin *right*, and on the second of these twigs, the result is spin *left*. Thus, on any given twig of any branch there is a definite result (spin *up*, *down*, *right*, or *left*), but the idea of "what would have happened if one had chosen the 'other branch' " makes no sense, for both results (*up* or *down*, or *right* or *left*) occur on different branches. Thus the results on "the other" branch are not definite.

theorem. It is drawn from informal discussions of the Fundamental Physics Group at the Lawrence Berkeley Laboratory, under the direction and sponsorship of Dr. Elizabeth Rauscher. These discussions, in turn, were based primarily upon the work of Henry Stapp.

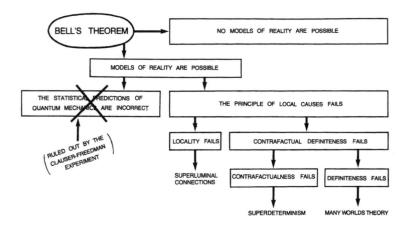

To summarize, Bell's theorem showed, in 1964, that either the statistical predictions of quantum theory are false or the principle of local causes is false. In 1972, Clauser and Freedman performed an experiment at Berkeley which validated the relevant statistical predictions of quantum theory. Therefore, according to Bell's theorem, the principle of local causes *must* be false.

The principle of local causes says that what happens in one area does not depend upon variables subject to the control of an experimenter in a distant space-like separated area. The simplest way to explain the failure of the principle of local causes is to conclude that what happens is one area *does* depend upon variables subject to the control of an experimenter in a distant space-like separated area. If this explanation is correct, then we live in a nonlocal universe ("locality fails") characterized by superluminal (faster than light) connections between apparently "separate parts."

However, there are other ways in which the principle of local causes can fail. The principle of local causes is based upon two tacit assumptions. The first tacit assumption is that we have the ability to determine our own actions, i.e., that we have a free will.* The second tacit assumption is that when we choose to do one thing in place of another, "what would have happened if . . ." would have produced definite results. These two assumptions together are what Stapp calls contrafactual definiteness.

If the first assumption (contrafactualness) fails, then we are led to a superdeterminism which precludes the idea of alternative possibilities. According to this type of determinism, it is not possible that the world *ever* could have been other than it is.

If the second assumption (definiteness) fails, then we are led to the Many Worlds theory in which the world continuously is splitting into separate and mutually inaccessible branches, each of which contains different editions of the same actors performing different acts at the same time on different stages which somehow are located in the same place.

There may be still ways to understand the failure of the principle of local causes, but the very fact that it must fail means that the world is in some way profoundly different from our ordinary ideas about it. (Perhaps we really are living in a dark cave.)

The "no models" option on the diagram is, in effect, the Copenhagen Interpretation of Quantum Mechanics (page 40). In 1927, the most famous assemblage of physicists in history decided that it might not ever be possible to construct a model of reality, i.e., to explain the way things "really are behind the scenes." Despite the tidal wave of "knowledge" which has swept over us for forty years, the Fundamental Physics Group found it necessary, like the physicists at Copenhagen a half century before them, to acknowledge that it might not be possible to construct a model of reality. This acknowledgment is more than

*Physicists usually express philosophical phrases (like "free will") in more precise terms. For example, the concept of free will is defined within this experimental situation as the tacit assumption that "each of the two observers, one located in area A, and the other located in area B, can choose between two possible observations [experiments]." These two choices are considered "free variables" in the context of the study of the observations made on the two-particle system.

a recognition of the limitations of this theory or that theory. It is a recognition emerging throughout the West that *knowledge itself* is limited. Said another way, it is a recognition of the difference between knowledge and wisdom.*

Classical science starts with the assumption of separate parts which together constitute physical reality. Since its inception, it has concerned itself with how these separate parts are related.

Newton's great work showed that the earth, the moon, and the planets are governed by the same laws as falling apples. The French mathematician Descartes invented a way of drawing pictures of relationships between different measurements of time and distance. This process (analytic geometry) is a wonderful tool for organizing a wealth of scattered data into one meaningful pattern. Herein lies the strength of western science. It brings huge tracts of apparently unrelated experience into a rational framework of simple concepts like the laws of motion. The starting point of this process is a mental attitude which initially perceives the physical world as fragmented and different experiences as logically unrelated. Newtonian science is the effort to find the relationships between pre-existing "separate parts."

Quantum mechanics is based upon the opposite epistemological assumption. Thus, there are profound differences between Newtonian mechanics and quantum theory.

*In fact, most physicists do *not* believe that it is worthwhile to think about these problems. The main thrust of the Copenhagen Interpretation, which is the interpretation of quantum theory accepted by the bulk of the scientific community, is that the proper goal of science is to provide a mathematical framework for organizing and expanding our experiences, rather than providing a picture of some reality that could lie behind these experiences. That is, most physicists today side with Bohr, rather than with Einstein, on the question of the utility of seeking a model of a reality that can be conceived of independently of our experience of it. From the Copenhagen point of view, quantum theory is satisfactory as it is, and the effort to "understand" it more deeply is not productive for science. Such efforts lead to perplexities of just the sort that we have been discussing. These perplexities seem to most physicists to be more philosophical than physical. Thus most physicists choose the "no model" option shown in the diagram on page 335.

The most fundamental difference between Newtonian physics and quantum mechanics is the fact that quantum mechanics is based upon *observations* ("measurements"). Without a measurement of some kind, quantum mechanics is mute. Quantum mechanics says nothing about what happens between measurements. In Heisenberg's words: "The term 'happens' is restricted to the observation."[15] This is very important, for it constitutes a philosophy of science unlike any before it.

We commonly say, for example, that we detect an electron at point A and then at point B, but strictly speaking, this is incorrect. According to quantum mechanics, there was no electron which traveled from point A to point B. There are only the measurements that we made at point A and at point B.

Quantum theory not only is closely bound to philosophy, but also—and this is becoming increasingly apparent—to theories of perception. As early as 1932, von Neumann explored this relation in his "Theory of Measurement." (Exactly when does the wave function associated with a particle collapse? When the particle strikes a photographic plate? When the photographic plate is developed? When the light rays from the developed photographic plate strike our retina? When the nerve impulses from the retina reach our brain?) (page 87).

Bohr's principle of complementarity (page 103) also addresses the underlying relation of physics to consciousness. The experimenter's choice of experiment determines which mutually exclusive aspect of the same phenomenon (wave or particle) will manifest itself. Likewise, Heisenberg's uncertainty principle (page 123) demonstrates that we cannot observe a phenomenon without changing it. The physical properties which we observe in the "eternal" world are enmeshed in our own perceptions not only psychologically, but ontologically as well.

The second most fundamental difference between Newtonian physics and quantum theory is that Newtonian physics predicts events and quantum mechanics predicts the probability of events. According to quantum mechanics, the only determinable relation between events is statistical—that is, a matter of probability.

David Bohm, Professor of Physics at Birkbeck College, University of London, proposes that quantum physics is, in fact, based upon a perception of a new order. According to Bohm, "We must turn physics around. Instead of starting with parts and showing how they work together (the Cartesian order) (page 24) we start with the whole."[16]

Bohm's theory is compatible with Bell's theorem. Bell's theorem implies that the apparently "separate parts" of the universe could be intimately connected at a deep and fundamental level. Bohm asserts that the most fundamental level is an *unbroken wholeness* which is, in his words, "that-which-is." All things, including space, time, and matter, are forms of that-which-is. There is an order which is enfolded into the very process of the universe, but that enfolded order may not be readily apparent.

For example, imagine a large hollow cylinder into which is placed a smaller cylinder. The space between the smaller cylinder and the larger cylinder is filled with a clear viscous liquid like glycerine (such a device actually exists).

Now suppose that we deposit a small droplet of ink on the surface of the glycerine. Because of the nature of the glycerine, the ink drop remains intact, a well-defined black spot floating on a clear liquid.

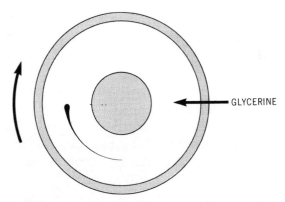

GLYCERINE

If we begin to rotate one of the cylinders, say in a clockwise direction, the drop of ink spreads out in the opposite direction, making a line which grows thinner and thinner until it disappears altogether. The ink droplet now is enfolded completely into the glycerine, *but it is still there*. When we rotate the cylinder in the opposite direction, the ink droplet reappears. A fine line appears which grows thicker and thicker and then collects into a single point.

If we continue the counterclockwise motion of the cylinder, the same thing happens, but in reverse. We can repeat this process as often as we like. Each time the ink spot becomes a fine line and disappears into the glycerine only to reappear again when the motion of the glycerine is reversed.

If it requires one complete revolution of the cylinder clockwise to make the droplet disappear completely, one complete revolution of the cylinder counterclockwise will make it reappear in its original shape and location. The number of revolutions required to make the droplet disappear or reappear is the enfolded order. Bohm calls this enfolded order the "implicate order," which means the same thing.

Suppose that we deposit a drop of ink on the surface of the glycerine, revolve the cylinder clockwise until the drop disappears (one revolution), add a second drop of ink to the glycerine, continue to revolve the cylinder in the same direction until *it* disappears (one more revolution), and then add a third drop of ink to the glycerine and revolve the cylinder one more revolution until the third drop also disappears. Now we have three ink drops enfolded into the glycerine. None of them are visible, but we know where each of them is in the implicate order.

When we revolve the cylinder in the opposite direction, one drop of ink (the third) appears after one revolution, another drop of ink (the second) appears after the next revolution, and another drop of ink (the first) appears after the third revolution. This is the unfolded, or "explicate," order. The three ink droplets appear to be unrelated in the explicate (unfolded) order, but we know that they *are* related in the implicate (enfolded) order.

If we consider the condensation of ink droplets in this experiment as "particles," we have Bohm's hypothesis of apparently random subatomic phenomena. "Particles" may appear in different places yet be connected in the implicate order. In Bohm's words, "Particles may be discontiguous in space (the explicate order) but contiguous in the implicate order."[17]

"Matter is a form of the implicate order as a vortex is a form of the water—it is not reducible to smaller particles."[18] Like "matter" and everything else, particles are forms of the implicate order. If this is difficult to grasp, it is because our minds demand to know, "What is the 'implicate order' the implicate order of?"

The "implicate order" is the implicate order of that-which-is. However, that-which-is *is* the implicate order. This world view is so different from the one that we are using that, as Bohm points out, "Description is totally incompatible with what we want to say."[19] Description is incompatible with what we want to say because our thinking is based upon an ancient Greek mode of thought. According to this mode of thought, only Being *is*. Therefore, Nonbeing *is not*. This way of thinking gives us a practical tool for dealing with the world, but it doesn't describe what happens. Actually, Nonbeing also *is*. Both Being and Nonbeing are that-which-is. *Everything*, even "emptiness," is that-which-is. There is nothing which is not that-which-is.

This way of looking at reality raises the question of the consciousness of the observer. Our minds demand to know, "What is the 'implicate order' the implicate order of?" because our culture has taught us to perceive only the explicate order (the Cartesian view). "Things" to us are intrinsically separate.

Bohm's physics require, in his words, a new "instrument of thought." A new instrument of thought such as is needed to understand Bohm's physics, however, would radically alter the consciousness of the observer, reorientating it toward a perception of the "unbroken wholeness" of which everything is a form.

However, such a perception would not cause an inability to see the explicate order. Bohm's physics contains an element of relativity parallel to that of Einstein's theories. The implicate or explicate nature of order (or order of nature) depends upon the perspective of the viewer. The problem is that our present viewpoint is limited to the perspective of the explicate order. From the perspective of the implicate order the apparently "separate elements" of the explicate order are intimately related. Even the phrases "elements" and "intimately related" imply a Cartesian separateness which does not exist. At the fundamental level of that-which-is, the "separate elements" which are "intimately related in the implicate order" *are* the implicate order.

The requirement for a new instrument of thought upon which to base Bohm's physics may not be as much of an obstacle as it first appears. There already exists an instrument of thought based upon an "unbroken wholeness." Furthermore, there exist a number of sophisticated psychologies, distilled from two thousand years of practice and introspection, whose sole purpose is to develop this thought instrument.

These psychologies are what we commonly call "Eastern religions." "Eastern religions" differ considerably among themselves. It would be a mistake to equate Hinduism, for example, with Buddhism, even though they are more like each other than either one of them is like a religion of the West. Nonetheless, all eastern religions (psychologies) are compatible in a very fundamental way with Bohm's physics and philosophy. All of them are based upon the experience of a pure, undifferentiated reality which is that-which-is.

While it would be naive to overstate the similarities between Bohm's physics and eastern philosophies, it would be foolish to ignore them. Consider, for example, the following sentences:

> The word "reality" is derived from the roots "thing" (*res*) and "think" (*revi*). "Reality" means "everything you can think about." This is not "that-which-is." No idea can capture "truth" in the sense of that-which-is.

> The ultimate perception does not originate in the brain or any material structure, although a material structure is necessary to manifest it. The subtle mechanism of knowing the truth does not originate in the brain.

> There is a similarity between thought and matter. All matter, including ourselves, is determined by "information." "Information" is what determines space and time.

Taken out of context, there is no absolute way of knowing whether these statements were made by Professor Bohm or a Tibetan Buddhist. In fact, these sentences were excerpted from different parts of two *physics* lectures that Professor Bohm gave at Berkeley in April, 1977. The first lecture was given on the campus to physics students. The second lecture was given in the Lawrence Berkeley Laboratory to a group of professional physicists. Two of these three statements were taken from the *second* lecture, the one given to the advanced physicists.

It is ironic that while Bohm's theories are received with some skepticism by most professional physicists, they would find an immediately sympathetic reception among the thousands of people in our culture who have turned their backs on science in their own quest for the ultimate nature of reality.

If Bohm's physics, or one similar to it, should become the main thrust of physics in the future, the dances of East and West could blend in exquisite harmony. Physics curricula of the twenty-first century could include classes in meditation.

The function of eastern religions (psychologies) is to allow the mind to escape the confines of the symbolic. According to this view, *everything* is a symbol, not only words and concepts, but also people and things. Beyond the confines of the symbolic lies that which is, pure awareness, the experience of the "suchness" of reality.

Nonetheless, every eastern religion resorts to the use of symbols to escape the realm of the symbolic. Some disciplines use symbols more than others, but all of them use symbols in one form or other. Therefore, the question arises, if pure awareness is considered distinct from the content of awareness, in what ways specifically does the content of awareness affect the realization of pure awareness? What types of content prompt the mind to leap forward? What enables it to activate the self-fulfilling capability to transcend itself?

It is very difficult to answer this question. Any answer is only a point of view. A point of view itself is limiting. To "understand" something is to give up some other way of conceiving it. This is another way of saying that the mind deals in forms of limitation. Nonetheless, there *is* a relationship between the content of awareness and the ability of the mind to transcend itself.

"Reality" is what we take to be true. What we take to be true is what we believe. What we believe is based upon our perceptions. What we perceive depends upon what we look for. What we look for depends upon what we think. What we think depends upon what we perceive. What we perceive determines what we believe. What we believe determines what we take to be true. What we take to be true is our reality.

The central focus of this process, initially at any rate, is "What we think." We at least can say that allegiance to a symbol of openness (Christ, Buddha, Krishna, "the infinite diversity of nature," etc.) seems to open the mind and that an open mind is often the first step in the process of enlightenment.

The psychological gestalt of physics has shifted radically in the last century to one of extreme openness. In the middle 1800s, Newtonian mechanics was at its zenith. There seemed to be no phenomenon which could not be explained in terms of mechanical models. All mechanical models were subject to long-established principles. The chairman of the physics department at Harvard discouraged graduate study because so few important matters remained unsolved.[20]

In a speech to the Royal Institution in 1900, Lord Kelvin reflected that there were only two "clouds" on the horizon of physics, the problem of black-body radiation and the Michelson-Morley experiment.[21] There was no doubt, said Kelvin, that they soon would be gone. He was wrong. Kelvin's two "clouds" signaled the end of the era that began with Galileo and Newton. The problem of black-body radiation led to Planck's discovery of the quantum of action. Within thirty years the entirety of Newtonian physics became a special limiting case of the newly developing quantum theory. The Michelson-Morley experiment foreshadowed Einstein's famous theories of relativity. By 1927, the foundations of the new physics, quantum mechanics and relativity, were in place.

In contrast to Kelvin's time, the allegiance of physicists today is to a symbol of extreme openness. Isidor Rabi, Nobel Prize winner and Chairman Emeritus of the Physics Department at Columbia University, wrote in 1975:

> I don't think that physics will ever have an end. I think that the novelty of nature is such that its variety will be infinite—not just in changing forms but in the profundity of insight and the newness of ideas . . .[22]

Stapp wrote in 1971:

> . . . human inquiry can continue indefinitely to yield important new truths.[23]

The "What we think" of physicists today is that the physics of nature, like human experience itself, is infinitely diverse.

Eastern religions have nothing to say about physics, but they have a great deal to say about human experience. In Hindu mythology, Kali, the Divine Mother, is the symbol for the infinite diversity of experience. Kali represents the entire physical plane. She is the drama, tragedy, humor, and sorrow of life. She is the brother, father, sister, mother, lover, and friend. She is the fiend, monster, beast, and brute. She is the sun and the ocean. She is the grass and the dew. She is our sense of accomplishment and our sense of doing worthwhile. Our thrill of discovery is a pendant on her bracelet. Our gratification is a spot of color on her cheek. Our sense of importance is the bell on her toe.

This full and seductive, terrible and wonderful earth mother *always* has something to offer. Hindus know the impossibility of seducing her or conquering her and the futility of loving her or hating her; so they do the only thing that they can do. They simply honor her.

In a particular story, Kali, the Divine Mother, is Sita, the wife of God. Ram is God. Ram, Sita, and Laksaman, who is Ram's brother, are walking along a jungle trail. The path is so narrow that most of the time Laksaman can see only Sita, who walks between him and

Ram. Every so often, however, the path turns in such a way that Laksaman can see his brother, God.

These powerful metaphors have application to the developing drama of physics. Although most physicists have little patience (professionally) with metaphors, physics itself has become a powerful metaphor. Twentieth-century physics is the story of a journey from intellectual entrenchment to intellectual openness, despite the conservative prove-it-to-me nature of individual physicists. The realization that the discoveries of physics *never* will end has brought physicists, as well as those who have followed the story of physics, to an extremely fertile plateau. This realization invites the intellect to leap forward, although at great risk to its present hegemony.

The Wu Li Masters know that physicists are doing more than "discovering the endless diversity of nature." They are dancing with Kali, the Divine Mother of Hindu mythology.

Buddhism is both a philosophy and a practice. Buddhist philosophy is rich and profound. Buddhist practice is called *Tantra. Tantra* is the Sanskrit word meaning "to weave." There is little that can be said about *Tantra*. It must be done.

Buddhist philosophy reached its ultimate development in the second century A.D. No one has been able to improve much on it since then. The distinction between Buddhist philosophy and *Tantra* is well defined. Buddhist philosophy can be intellectualized. *Tantra* cannot. Buddhist philosophy is a function of the rational mind. *Tantra* transcends rationality. The most profound thinkers of the Indian civilization discovered that words and concepts could take them only so far. Beyond that point came the actual doing of a practice, the experience of which was ineffable. This did not prevent them from progressively refining the practice into an extremely effective and sophisticated set of techniques, but it did prevent them from being able to describe the experiences which these techniques produce.

The practice of *Tantra* does not mean the end of rational thought. It means the integration of thought based on symbols into larger spectrums of awareness. (Enlightened people still remember their zip codes.)

The development of Buddhism in India shows that a profound and penetrating intellectual quest into the ultimate nature of reality can culminate in, or at least set the stage for, a quantum leap beyond rationality. In fact, on an individual level, this is one of the roads to enlightenment. Tibetan Buddhism calls it the Path without Form, or the Practice of Mind. The Path without Form is prescribed for people of intellectual temperament. The science of physics is following a similar path.

The development of physics in the twentieth century already has transformed the consciousness of those involved with it. The study of complementarity (page 103), the uncertainty principle (page 123), quantum field theory (page 222), and the Copenhagen Interpretation of Quantum Mechanics (page 40) produces insights into the nature of reality very similar to those produced by the study of eastern philosophy. The profound physicists of this century increasingly have become aware that they are confronting the ineffable.

Max Planck, the father of quantum mechanics, wrote:

> Science . . . means unresting endeavor and continually progressing development toward an aim which the poetic intuition may apprehend, but which the intellect can never fully grasp.[24]

We are approaching the end of science. "The end of science" does not mean the end of the "unresting endeavor and continually progressing development" of more and more comprehensive and useful physical theories. (Enlightened physicists remember their zip codes, too.) The "end of science" means the coming of western civilization, in its own time and in its own way, into the higher dimensions of human experience.

Professor G. F. Chew, Chairman of the Physics Department at Berkeley, remarked, in reference to a theory of particle physics:

> Our current struggle [with certain aspects of advanced physics] may thus be only a foretaste of a completely new form of

human intellectual endeavor, one that will not only lie outside physics but will not even be describable as "scientific."[25]

We need not make a pilgrimage to India or Tibet. There is much to learn there, but here at home, in the most inconceivable of places, amidst the particle accelerators and computers, our own Path without Form is emerging.

Al Huang, the Tai Chi Master who created the metaphor of Wu Li, once wrote ". . . sooner or later we reach a dead end when we talk."[26] He could as well have said that sooner or later we go round in circles when we talk since going round in a circle is one kind of dead end.

As we sat in a cabin at Esalen and talked late into the night, my new friend, David Finkelstein, spoke to us softly.

> I think it would be misleading to call particles the entities involved in the most primitive events of the theory [quantum topology] because they don't move in space and time, they don't carry mass, they don't have charge, they don't have energy in the usual sense of the word.

QUESTION: So what is it that makes events at that level?

ANSWER: Who are the dancers and who the dance? They have no attributes other than the dance.

QUESTION: What is "they"?

ANSWER: The things that dance, the dancers. My God! We're back to the title of the book.[27]

Notes

Epigraphs

1. Albert Einstein and Leopold Infeld, *The Evolution of Physics*, New York, Simon and Schuster, 1938, p. 27.
2. Werner Heisenberg, *Physics and Philosophy*, Harper Torchbooks, New York, Harper & Row, 1958, p. 168.
3. Erwin Schrödinger, *Science and Humanism*, Cambridge, England, Cambridge University Press, 1951, pp. 7–8.

Big Week at Big Sur (pp. 3–18)

1. Al Chung-liang Huang, *Embrace Tiger, Return to Mountain*, Moab, Utah, Real People Press, 1973, p. 1.
2. Albert Einstein and Leopold Infeld, *The Evolution of Physics*, New York, Simon and Schuster, 1938, p. 31.
3. Isidor Rabi, "Profiles—Physicists, I," *The New Yorker Magazine*, October 13, 1975.

Einstein Doesn't Like It (pp. 19–45)

1. Albert Einstein and Leopold Infeld, *The Evolution of Physics*, New York, Simon and Schuster, 1961, p. 31.

2. *Ibid.*, p. 152.

3. Werner Heisenberg, *Across the Frontiers,* New York, Harper & Row, 1974, p. 114.

4. Isaac Newton, *Philosophiae Naturalis Principia Mathematica* (trans. Andrew Motte), reprinted in *Sir Isaac Newton's Mathematical Principles of Natural Philosophy and His System of the World* (revised trans. Florian Cajori), Berkeley, University of California Press, 1946, p. 547.

5. *Proceedings of the Royal Society of London,* vol. 54, 1893, p. 381, which refers to *Correspondence of R. Bentley,* vol. 1, p. 70. There is also a discussion of action-at-a-distance in a lecture of Clerk Maxwell in *Nature,* vol. VII, 1872, p. 325.

6. Joseph Weizenbaum, *Computer Power and Human Reason,* San Francisco, Freeman, 1976.

7. Niels Bohr, *Atomic Theory and Human Knowledge,* New York, John Wiley, 1958, p. 62.

8. J. A. Wheeler, K. S. Thorne, and C. Misner, *Gravitation,* San Francisco, Freeman, p. 1273.

9. Carl G. Jung, *Collected Works,* vol. 9, Bollingen Series XX, Princeton, Princeton University Press, 1969, pp. 70–71.

10. Carl G. Jung and Wolfgang Pauli, *The Interpretation of Nature and the Psyche,* Bollingen Series LI, Princeton, Princeton University Press, 1955, p. 175.

11. Albert Einstein, "On Physical Reality," *Franklin Institute Journal,* 221, 1936, 349ff.

12. Henry Stapp, "The Copenhagen Interpretation and the Nature of Space-Time," *American Journal of Physics,* 40, 1972, 1098ff.

13. Robert Ornstein, ed., *The Nature of Human Consciousness,* New York, Viking, 1974, pp. 61–149.

Living? (pp. 49–73)

1. Victor Guillemin, *The Story of Quantum Mechanics,* New York, Scribner's, 1968, pp. 50–51.

2. Max Planck, *The Philosophy of Physics,* New York, Norton, 1936, p. 59.

3. Henry Stapp, "Are Superluminal Connections Necessary?" *Nuovo Cimento,* 40B, 1977, 191.
4. Evan H. Walker, "The Nature of Consciousness," *Mathematical Biosciences,* 7, 1970, 175–76.
5. Werner Heisenberg, *Physics and Philosophy,* New York, Harper & Row, 1958, p. 41.

What Happens (pp. 74–97)

1. Max Born and Albert Einstein, *The Born-Einstein Letters,* New York, Walker and Company, 1971, p. 91. (The precise wording of this statement varies somewhat from translation to translation. This is the version popularly attributed to Einstein.)
2. Henry Stapp, "S-Matrix Interpretation of Quantum Theory," Lawrence Berkeley Laboratory preprint, June 22, 1970 (revised edition: *Physical Review,* D3, 1971, 1303ff).
3. *Ibid.*
4. *Ibid.*
5. Werner Heisenberg, *Physics and Philosophy,* Harper Torchbooks, New York, Harper & Row, 1958, p. 41.
6. Henry Stapp, "Mind, Matter, and Quantum Mechanics," unpublished paper.
7. Hugh Everett III, " 'Relative State' Formulation of Quantum Mechanics," *Reviews of Modern Physics,* vol. 29, no. 3, 1957, pp. 452–62.

The Role of "I" (pp. 101–127)

1. Niels Bohr, *Atomic Theory and the Description of Nature,* Cambridge, England, Cambridge University Press, 1934, p. 53.
2. Werner Heisenberg, *Physics and Philosophy,* Harper Torchbooks, New York, Harper & Row, 1958, p. 42.
3. Werner Heisenberg, *Across the Frontiers,* New York, Harper & Row, 1974, p. 75.
4. Erwin Schrödinger, "Image of Matter," in *On Modern Physics,* with W. Heisenberg, M. Born, and P. Auger, New York, Clarkson Potter, 1961, p. 50.

5. Max Born, *Atomic Physics*, New York, Hafner, 1957, p. 95.

6. *Ibid.*, p. 96.

7. *Ibid.*, p. 102.

8. Werner Heisenberg, *Physics and Beyond*, New York, Harper & Row, 1971, p. 76.

9. Niels Bohr, *Atomic Theory and Human Knowledge*, New York, John Wiley, 1958, p. 60.

10. Born, *op. cit.*, p. 97.

11. Heisenberg, *Physics and Philosophy*, *op. cit.*, p. 58.

Beginner's Mind (pp. 131–149)

1. Shunryu Suzuki, *Zen Mind, Beginner's Mind*, New York, Weatherhill, 1970, pp. 13–14.

2. Henry Miller, "Reflections on Writing," in *Wisdom of the Heart*, Norfolk, Connecticut, New Directions Press, 1941 (reprinted in *The Creative Process*, by B. Ghiselin (ed.), Berkeley, University of California Press, 1954, p. 186).

3. KQED Television press conference, San Francisco, California, December 3, 1965.

4. Werner Heisenberg, *Physics and Philosophy*, Harper Torchbooks, New York, Harper & Row, 1958, p. 33.

Special Nonsense (pp. 150–178)

1. Albert Einstein, "Aether und Relativitätstheorie," 1920, trans. W. Perret and G. B. Jeffery, *Side Lights on Relativity*, London, Methuen, 1922 (reprinted in *Physical Thought from the Presocratics to the Quantum Physicists* by Shmuel Sambursky, New York, Pica Press, 1975, p. 497).

2. *Ibid.*

3. *Ibid.*

4. Albert Einstein, "Die Grundlage der Allgemeinen Relativitätstheorie," 1916, trans. W. Perret and G. B. Jeffery, *Side Lights on Relativity*, London, Methuen, 1922 (reprinted in *Physical Thought from the Presocratics to the Quantum Physicists* by Shmuel Sambursky, New York, Pica Press, 1975, p. 491).

5. Einstein, "Aether und Relativitätstheorie," *op. cit.*, p. 496.

6. J. Terrell, *Physical Review*, 116, 1959, 1041.

7. Isaac Newton, *Philosophiae Naturalis Principia Mathematica* (trans. Andrew Motte), reprinted in *Sir Isaac Newton's Mathematical Principles of Natural Philosophy and His System of the World* (revised trans. Florian Cajori), Berkeley, University of California Press, 1946, p. 6.

8. From "Space and Time," an address to the 80th Assembly of German Natural Scientists and Physicians, Cologne, Germany, September 21, 1908 (reprinted in *The Principles of Relativity*, by A. Lorentz, A. Einstein, H. Minkowski, and H. Weyle, New York, Dover, 1952, p. 75).

9. Albert Einstein and Leopold Infeld, *The Evolution of Physics*, New York, Simon and Schuster, 1961, p. 197.

General Nonsense (pp. 179–209)

1. Albert Einstein and Leopold Infeld, *The Evolution of Physics*, New York, Simon and Schuster, 1961, p. 197.

2. *Ibid.*, p. 219.

3. *Ibid.*, pp. 33–34.

4. David Finkelstein, "Past-Future Asymmetry of the Gravitational Field of a Point Particle," *Physical Review*, 110, 1958, 965.

The Particle Zoo (pp. 213–235)

1. Goethe, *Theory of Colours*, Pt. II (Historical), iv, 8 (trans. C. L. Eastlake, London, 1840; repr., M.I.T. Press, Cambridge, Massachusetts, 1970).

2. Werner Heisenberg, *Across the Frontiers*, New York, Harper & Row, 1974, p. 162.

3. Werner Heisenberg *et al.*, *On Modern Physics*, New York, Clarkson Potter, 1961, p. 13.

4. David Bohm, *Causality and Chance in Modern Physics*, Philadelphia, University of Pennsylvania Press, 1957, p. 90.

5. Werner Heisenberg, *Physics and Beyond*, New York, Harper & Row, 1971, p. 41.

6. Werner Heisenberg *et al.*, *On Modern Physics*, *op. cit.*, p. 34.
7. Victor Guillemin, *The Story of Quantum Mechanics*, New York, Scribner's, 1968, p. 135.
8. Max Born, *The Restless Universe*, New York, Dover, 1951, p. 206.
9. *Ibid.*
10. *Ibid.*
11. Kenneth Ford, *The World of Elementary Particles*, New York, Blaisdell, 1963, pp. 45–46.

The Dance (pp. 236–279)

1. Louis de Broglie, "A General Survey of the Scientific Work of Albert Einstein," in *Albert Einstein, Philosopher-Scientist*, vol. 1, Paul Schilpp (ed.), Harper Torchbooks, New York, Harper & Row, 1949, p. 114.
2. Richard Feynman, "Mathematical Formulation of the Quantum Theory of Electromagnetic Interaction," in Julian Schwinger (ed.) *Selected Papers on Quantum Electrodynamics* (Appendix B), New York, Dover, 1958, p. 272.
3. Kenneth Ford, *The World of Elementary Particles*, New York, Blaisdell, 1963, p. 208 and cover.
4. Sir Charles Eliot, *Japanese Buddhism*, New York, Barnes and Noble, 1969, pp. 109–10.

More Than Both (pp. 283–311)

1. John von Neumann, *The Mathematical Foundations of Quantum Mechanics* (trans. Robert T. Beyer), Princeton, Princeton University Press, 1955.
2. *Ibid.*, p. 253.
3. Werner Heisenberg, *Physics and Beyond*, New York, Harper & Row, 1971, p. 206.
4. Max Born, *Atomic Physics*, New York, Hafner, 1957, p. 97.
5. Transcribed from tapes recorded at the Esalen Conference on Physics and Consciousness, Big Sur, California, January 1976.
6. Albert Einstein, Boris Podolsky, and Nathan Rosen, "Can Quantum-

Mechanical Description of Physical Reality Be Considered Complete?" *Physical Review*, 47, 1935, 777ff.

7. Werner Heisenberg, *Across the Frontiers*, New York, Harper & Row, 1974, p. 72.

8. Esalen Tapes, *op. cit.*

9. Garrett Birkhoff and John von Neumann, "The Logic of Quantum Mechanics," *Annals of Mathematics*, vol. 37, 1936.

10. Esalen Tapes, *op. cit.*

The End of Science (pp. 312–348)

1. Longchenpa, "The Natural Freedom of Mind," trans. Herbert Guenther, *Crystal Mirror*, vol. 4, 1975, p. 125.

2. Albert Einstein, Boris Podolsky, and Nathan Rosen, "Can Quantum-Mechanical Description of Physical Reality Be Considered Complete?" *Physical Review*, 47, 1935, 777ff.

3. Erwin Schrödinger, "Discussions of Probability Relations between Separated Systems," *Proceedings of the Cambridge Philosophical Society*, 31, 1935, 555–62.

4. Albert Einstein, "Autobiographical Notes," in Paul Schilpp (ed.), *Albert Einstein, Philosopher-Scientist*, Harper Torchbooks, New York, Harper & Row, 1949, p. 85.

5. *Ibid.*, p. 87.

6. *Ibid.*, p. 85.

7. Henry Stapp, "S-Matrix Interpretation of Quantum Theory," Lawrence Berkeley Laboratory preprint, June 22, 1970 (revised edition: *Physical Review*, D3, 1971, 1303ff).

8. Stuart Freedman and John Clauser, "Experimental Test of Local Hidden Variable Theories," *Physical Review Letters*, 28, 1972, 938ff.

9. Henry Stapp, "Bell's Theorem and World Process," *Il Nuovo Cimento*, 29B, 1975, 271.

10. Alain Aspect, Jean Dalibard, and Gérard Roger, "Experimental Test of Bell's Inequalities Using Time-Varying Analyzers," *Physical Review Letters*, vol. 49, no. 25, 1982, 1804.

11. John Clauser and Abner Shimony, "Bell's Theorem: Experimental

Tests and Implications," *Rep Prog Phys,* vol 41, 1978, 1881; Bernard d'Espagnat, "The Quantum Theory and Reality," *Scientific American,* Nov. 1979.

12. Henry Stapp, "Are Superluminal Connections Necessary?" *Il Nuovo Cimento,* 40B, 1977, 191.

13. David Bohm and B. Hiley, "On the Intuitive Understanding of Non-locality as Implied by Quantum Theory" (preprint, Birkbeck College, University of London, 1974).

14. Henry Stapp, "S-Matrix Interpretation," *op. cit.*

15. Werner Heisenberg, *Physics and Philosophy,* Harper Torchbooks, New York, Harper & Row, 1958, p. 52.

16. Lecture given April 6, 1977, University of California at Berkeley.

17. *Ibid.*

18. *Ibid.*

19. *Ibid.*

20. Victor Guillemin, *The Story of Quantum Mechanics,* New York, Scribner's, 1968, p. 19.

21. Lord Kelvin (Sir William Thompson), "Nineteenth-Century Clouds over the Dynamical Theory of Heat and Light," *Philosophical Magazine,* 2, 1901, 1–40.

22. Isidor Rabi, "Profiles—Physicist, II," *The New Yorker Magazine,* October 20, 1975.

23. Henry Stapp, "The Copenhagen Interpretation and the Nature of Space-Time," *American Journal of Physics,* 40, 1972, 1098.

24. Max Planck, *The Philosophy of Physics,* New York, Norton, 1936, p. 83.

25. This quotation was given to the Fundamental Physics Group, Lawrence Berkeley Laboratory, November 21, 1975, (during an informal discussion of the bootstrap theory), by Dr. Chew's colleague, F. Capra.

26. Al Chung-liang Huang, *Embrace Tiger, Return to Mountain,* Moab, Utah, Real People Press, 1973, p. 14.

27. Transcribed from tapes recorded at the Esalen Conference on Physics and Consciousness, Big Sur, California, January 1976.

Bibliography

Barnett, L., *The Universe and Dr. Einstein*, New York, Harper & Row, 1948.

Birkhoff, G., and von Neumann, J., "The Logic of Quantum Mechanics," *Annals of Mathematics*, vol 37, no. 4, Oct. 1936.

Bohm, D., *Causality and Chance in Modern Physics*, Philadelphia, University of Pennsylvania Press, 1957.

Bohm, D., and Hiley, B., "On the Intuitive Understanding of Non-locality as Implied by Quantum Theory," (preprint, Birkbeck College, University of London, 1974).

Bohr, N., *Atomic Theory and Human Knowledge*, New York, John Wiley, 1958.

Bohr, N., *Atomic Theory and the Description of Nature*, Cambridge, England, Cambridge University Press, 1934.

Born, M., *Atomic Physics*, New York, Hafner, 1957.

Born, M., *The Restless Universe*, New York, Dover, 1951.

Born, M., and Einstein, A., *The Born-Einstein Letters* (trans. Irene Born), New York, Walker and Company, 1971.

Capra, F., *Chronology of the Development of Quantum Mechanics*, unpublished paper prepared for the Physics/Consciousness Research Group, J. Sarfatti, Ph.D., Director.

Capra, F., *The Tao of Physics*, Berkeley, Shambhala, 1975.

de Broglie, L., "A General Survey of the Scientific Work of Albert Einstein," in Schilpp, P. (ed.), *Albert Einstein, Philosopher-Scientist*, vol. 1, New York, Harper & Row, 1949, p. 114.

De Witt, and Graham, N., *The Many Worlds Interpretation of Quantum Mechanics*, Princeton, Princeton University Press, 1973.

Eddington, A., *The Mathematical Theory of Relativity*, Cambridge, England, Cambridge University Press, 1923.

Eddington, A., *Space, Time, and Gravitation*, Cambridge, England, Cambridge University Press, 1920.

Einstein, A., "Aether und Relativitätstheorie," 1920 (trans. Perret, W., and Jeffery, G., *Side Lights on Relativity*, London, Methuen, 1922).

Einstein, A., "Autobiographical Notes," in Schilpp, P. (ed.), *Albert Einstein, Philosopher-Scientist*, vol. 1, New York, Harper & Row, 1959, p. 1ff.

Einstein, A., "Die Grundlage de Allgemeinen Relativitätstheorie," 1916 (trans. Perret, W., and Jeffery, G., *Side Lights on Relativity*, London, Metheun, 1922).

Einstein, A., "On Physical Reality," *Franklin Institute Journal*, 221, 1936, 349ff.

Einstein, A., and Infeld, L., *The Evolution of Physics*, New York, Simon and Schuster, 1961.

Einstein, A., Podolsky, B., and Rosen, N., "Can Quantum-Mechanical Description of Physical Reality Be Considered Complete?" *Physical Review*, 47, 1935, 777.

Eliot, C., *Japanese Buddhism*, New York, Barnes and Noble, 1969.

Feynman, R., "Mathematical Formulation of the Quantum Theory of Electromagnetic Interaction," in Schwinger, J. (ed.), *Selected Papers of Quantum Electrodynamics*, New York, Dover, 1958, p. 272ff.

Finkelstein, D., "Beneath Time: Explorations in Quantum Topology," unpublished paper.

Finkelstein, D., "Past-Future Asymmetry of the Gravitational Field of a Point Particle," *Physical Review*, 110, 1958, 965.

Ford, K., *The World of Elementary Particles*, New York, Blaisdell, 1963.

Freedman, S., and Clauser, J., "Experimental Test of Local Hidden Variable Theories," *Physical Review Letters,* 28, 1972, 938.

Goethe, *Theory of Colours,* trans. Eastlake, C. L., London, 1840: repr. M.I.T. Press, Cambridge, Mass., 1970.

Guillemin, V., *The Story of Quantum Mechanics,* New York, Scribner's, 1968.

Hafele, J., and Keating, R., *Science* 177, 1972.

Hawking, S. W., "Singularities in the Geometry of Space-time," Adams Prize, Cambridge University, 1966.

Heisenberg, W., *Across the Frontiers,* New York, Harper & Row, 1974.

Heisenberg, W., *Physics and Beyond,* New York, Harper & Row, 1971.

Heisenberg, W., *Physics and Philosophy,* New York, Harper & Row, 1958.

Heisenberg, W., *et al., On Modern Physics,* New York, Clarkson Potter, 1961.

Herbert, N., "More Than Both: A Key to Quantum Logic," unpublished paper (available from C-Life Institute, Box 261, Boulder Creek, Cal. 95006).

Herbert N., "Thru the Looking Glass: Alice's Analysis of Quantum Logic," unpublished paper (available from C-Life Institute, Box, 261, Boulder Creek, Cal. 95006).

Herbert, N., "Where Do Parts Come From?" unpublished paper (available from C-Life Institute, Box 261, Boulder Creek, Cal. 95006).

Huang, A., *Embrace Tiger, Return to Mountain,* Moab, Utah, Real People Press, 1973.

Jung, C., *Collected Works,* vol. 9 (Bollingen Series XX), Princeton, Princeton University Press, 1969.

Jung, C., and Pauli, W., *The Interpretation of Nature and the Psyche* (Bollingen Series LI), Princeton, Princeton University Press, 1955.

Kelvin, Lord (Sir William Thompson), "Nineteenth-Century Clouds over the Dynamical Theory of Heat and Light," *Philosophical Magazine,* 2, 1901, 1–40.

Longchenpa, "The Natural Freedom of Mind," trans. Guenther, H., in *Crystal Mirror,* vol. 4, 1975, p. 125.

Lorentz, A., *et al., The Principle of Relativity,* New York, Dover, 1952.

Miller, H., "Reflections on Writing," *Wisdom of the Heart,* Norfolk,

Conn., New Directions, 1941 (repr. in Ghiselin [ed.], *The Creative Process,* Berkeley, University of California Press, 1954).

Murchie, G., *Music of the Spheres,* vols. 1 and 2, New York, Dover, 1961.

Newton, I., *Philosophiae Naturalis Principia Mathematica* (trans. Andrew Motte), reprinted in *Sir Isaac Newton's Mathematical Principles of Natural Philosophy and His System of the World* (revised trans. Florian Cajori), Berkeley, University of California Press, 1946.

Oppenheimer, J. R., and Snyder, H., "On Continual Gravitational Contraction," *Physical Review,* 56, 1939, 455–59.

Ornstein, R. (ed.), *The Nature of Human Consciousness,* New York, Viking, 1974.

Penrose, R., "Gravitational Collapse and Space-time Singularities," *Physical Review Letters,* 14, 1965, 57–59.

Planck, M., "Neue Bahnen der physikalischen Erkenntnis," 1913 (trans. D'Albe, F., *Phil. Mag.,* vol. 28, 1914).

Planck, M., *The Philosophy of Physics,* New York, Norton, 1936.

Rabi, I., "Profiles—Physicist, I," *The New Yorker Magazine,* October 13, 1975.

Rabi, I., "Profiles—Physicist, II," *The New Yorker Magazine,* October 20, 1975.

Russell, B., *The ABC of Relativity,* London, George Allen & Unwin, 1958.

Sambursky, S., *Physical Thought from the Presocratics to the Quantum Physicists,* New York, Pica, 1975.

Sarfatti, J., "The Case for Superluminal Information Transfer," *MIT Technology Review,* vol. 79, no. 5, 1977, p. 3ff.

Sarfatti, J., "Mind, Matter, and Einstein," unpublished paper.

Sarfatti, J., "The Physical Roots of Consciousness," in Mishlove, J., *The Roots of Consciousness,* New York, Random House, 1975, pp. 279ff.

Sarfatti, J., "Reply to Bohm-Hiley," *Psychoenergetic Systems,* London, Gordon & Breach, vol. 2, 1976, pp. 1–8.

Schilpp, P., *Albert Einstein, Philosopher-Scientist,* vol. 1, New York, Harper & Row, 1949.

Schrödinger, E., "Discussions of Probability Relations Between Separated Systems," *Proceedings of the Cambridge Philosophical Society,* vol. 31, 1935.

Schrödinger, E., "Image of Matter," in Heisenberg, W., et. al., *On Modern Physics,* New York, Clarkson Potter, 1961, pp. 50ff.

Shamos, M. (ed.), *Great Experiments in Physics,* New York, Holt-Dryden, 1959.

Stapp, H., "Are Superluminal Connections Necessary?" *Il Nuovo Cimento,* 40B, 1977, 191.

Stapp, H., "Bell's Theorem and World Process," *Il Nuovo Cimento,* 29B, 1975, 270.

Stapp, H., "The Copenhagen Interpretation and the Nature of Space-Time," *American Journal of Physics,* 40, 1972, 1098.

Stapp, H., "S-Matrix Interpretation of Quantum Theory," *Physical Review,* D3, 1971, 1303.

Stapp, H., "Theory of Reality," *Foundations of Physics,* 7, 1977, 313.

Suzuki, S., *Zen Mind, Beginner's Mind,* New York, Weatherhill, 1970.

Targ, R., and Puthoff, H., *Mind-Reach,* New York, Delacorte Press, 1977.

Taylor, J., *Black Holes: The End of the Universe?* New York, Random House, 1973.

Terrell, J., *Physical Review,* 116, 1959, 1041.

Von Neumann, J., *The Mathematical Foundations of Quantum Mechanics,* trans. Beyer, R., Princeton, Princeton University Press, 1955.

Walker, E., "The Nature of Consciousness," *Mathematical Biosciences,* 7, 1970.

Weisskopf, V., *Physics in the Twentieth Century,* Cambridge, Mass., M.I.T. Press, 1972.

Weizenbaum, J., *Computer Power and Human Reason,* San Francisco, Freeman, 1976.

Wheeler, J., et al., *Gravitation,* San Francisco, Freeman, 1973.

Witten (ed.), *Gravitation: An Introduction to Current Research,* New York, Wiley, 1962.

Index

Absolute motion, 142, 183
Absolute non-motion, 141–47,
    151–53, 184, 209
Absolute time, 166
Absolute truth, 41–42
Acceleration, 178
    gravity and, 186
    negative, 186–87
    positive, 186–87
Accelerators
    electron, 226
    feeder, 226
    particle, 215, 217–18, 219, 226
Admissions, 303
Albert Einstein, Philosopher-Scientist
    (ed. Schilpp), 262*n*
Allowed transitions, 302
Amplitude, 61
Analytic geometry, 337
Anderson, Carl, 235
Angular momentum, 231–34
Antimuon, 330
Anti-particles, 235, 239–40, 242,
    243, 275, 276
Anti-protons, 263, 264
Aristotelianism, 72
Aristotle, 23, 287
Aspect, Alain, xxi, 327–28
Astronauts, 23
Astronomy, 53, 201–204
    Ptolemaic, 217*n*
Atomic bombs, 175
Atomic level, 34

Atomic oscillators, 53–54, 55
Atomic phenomena, 21
Atomic reactors, 175
Atoms, 53–54, 114
    Bohr's theory on, 14–16, 114,
        119
    construction of, 12–13
    ground state of, 14–15
    the new physics and, 119–20
"Auguries of Innocence" (Blake),
    167
Authentic power, xxix–xxx
Avatamsaka Sutra, 266*n*
Awareness movement, 173

Baryon number, 234*n*
Baryons, 228, 234, 235, 270–71,
    273
Be Here Now (Dass), 173
Being, Non-being and, 341
Bell, J. S., xxi, 75, 313
Bell Telephone Laboratories, 107
Bell's theorem, 252*n*, 285, 313–14,
    322–26, 329–33, 339
    quantum theory and, 322–23,
        325–26, 329, 331, 334–35
"Beneath Time" (Finkelstein), 311
Bently, Richard, 25
Bible, 97, 173–74
Big Bang theory, 329*n*
Binary star system, 207
Birkbeck College, 221, 339
Birkhoff, Garrett, 292–93, 302, 310,
    311

Black-body radiation, 54–55, 106, 233, 344

Black hole singularity, 206–207

Black hole theory, 205–209

Blake, William, 167

Blofeld, John, 266n

Bohm, David, xxi, 221, 316, 327, 329–30, 339–43

Bohr, Niels, xix, xx, xxv, 14–16, 28, 38, 40n, 41n, 57, 72, 109, 118, 125, 224
   complementarity and, 41n, 103, 106, 223–24n, 338, 347
   Einstein and, 290
   Planck's constant and, 233
   planetary model of the atom, 110, 114–15, 117, 119–20, 223
   on quantum mechanics, 28, 125–26, 289, 337n
   theory on atoms, 14–16

Boltzmann's constant, 58n

Born, Max, xx, 75, 117–18, 126, 232, 289

Bound, greatest lower, 308

Boyle, Robert, 36

Boyle's Law, 36

Bubble-chamber physics, 218n

Bubble chambers, 215, 218, 221, 224, 251

Buddha, 264, 344

Buddhism, xxxiv, 88, 91, 174, 201, 229, 241, 262–63, 264–66, 267–68, 312, 342–43, 346–47
   Mahayana, 266, 267–68
   particle physics and, 262–63, 264–66
   physics and, 311
   Tantric, 220–21, 346–47
   Tibetan, 312, 343, 347
   Zen, 132, 229

Buddhist Text Translation Society, 266n

Calculus, logical, 288

Cameras, computer-triggered, 218

"Can Quantum-Mechanical Description of Physical Reality Be Considered Complete?" (Einstein, Podolsky, and Rosen), 317

Carbon, 120

Causality, 43, 70, 125–26, 286–87

Causes, local, principle of, 320, 325, 329, 331–36

CERN, 313

Chain reactions, 175

Charge, 234n, 335, 270

Charm, 217n, 234n

Chemical reactions, human reactions and, 50–51

Chew, Professor G. F., 266n, 347

Chlorine, 50

Christ, 344

Church, the
   Newton and, 23–24, 30
   physics and, 24

Circles, 190, 193–96, 198
   circumferences, 194–96
   radii, 193–96

Circumferences, 194–95

Classical logic, 291–93, 298–99, 302, 306–307, 311
   subatomic particles and, 291–93

Classical physics, xxv, 53–54, 150–51, 289–90
   quantum mechanics and, 94–96, 125

Classical transformation laws, 140, 152, 153–54, 164–65

Clauser, John, xxi 326, 329, 332, 335

Clauser-Freedman experiment, xxi 323n, 326–27, 329

Coherent superposition, 299–300

Collapse of the wave function, 82–85, 95–96, 330–31n

Color, Planck's constant and, 56–57

Communications, superluminal, 326–28

Complementarity, 41n, 103, 105, 106, 223–24n, 338, 347

Compton, Authur, 103–105

Compton scattering, 104–105
Computers, 26, 215, 217
Computer-triggered cameras, 218
Confucius, 169
Congruence, 180*n*
Consciousness, physics and, xxviii, xxx, 88, 92–93, 96
Conservation laws, 175–77, 217, 269–71
  subatomic particles and, 268–71
Conservation laws of family numbers, 270
Conservation laws of mass-energy, 175, 176, 248, 249, 255, 261–62, 269, 270
Constant of proportionality, 63
Continuums, 167–68
Contrafactual definiteness, 332–33, 336
Co-ordinate systems, 138–41, 180–82, 187*n*, 196–97, 202
  inertial, 138–41, 182–84, 186, 194, 198
  non-inertial, 196
Copenhagen Interpretation of Quantum Mechanics, xx, 40–44, 76, 88, 92, 93, 94–95, 317, 336, 347
Copernicus, Nicolaus, 101, 147, 213–15, 217*n*
Correlation, 78–79
cosmic radiation, 158
Counter physics, 218*n*
cummings, e. e., 291
Cyclotrons, 236
Cygnus X-1, 207

D particles, 228*n*
Dass, Ram, 173
Davisson, Clinton, xx, 107
Davisson-Germer experiment, 107, 109
De Broglie, Louis, xx, 106–107, 109, 110, 114, 115, 117–18, 122, 233, 245
De Broglie's equation, 107*n*, 110
Decay times, 240

Deceleration, 178
Definiteness, 336
  contrafactual, 332–33, 336
Deflection, starlight, 203–204
Descartes, René, 12, 24, 26, 141, 337
Detectors, photon, 82. *See also* Photomultiplier tubes
Determinism, 37, 333, 336
Diffraction, 63–66, 134
  electron, 107–108
Dirac, Paul, xx, 103*n*, 222, 234–35, 239, 251, 263
Discontinuous structure of nature, 52–53
Distributivity, law of, 292–93, 307–10
Doppler effect, 144*n*
Double exposure, 299
Double-slit experiment, 66–87, 92, 105, 110, 314
  See also Young's double-slit experiment
Doublets, 306–307
Downhill interactions, 269
Duration, 26
Dylan, Bob, 133

Earth
  the moon and, 27
  motion and, 137–39, 145–48
Eastern literature
  quantum mechanics and, 101
  relativity and, 101
Eastern religions, 342–48
Eddington, Arthur, 189, 192*n*
Einstein, Albert, xix, xx, xxiv, xxv, 9, 25, 57–60, 118, 121, 142, 216, 290, 316–21, 341
  Bohr and, 290
  general theory of relativity, 25, 94, 134, 179–209, 290, 311, 344
    astronomy and, 201–205
    black hole theory, 205–209
    equations of, 200
    gravity and acceleration, 186–209

Newtonian physics and, 203–204
principle of equivalence, 183, 186, 187
photon theory of light, 72, 104, 106–107, 110, 133–34
Planck and, 59, 72
Planck's constant, 233
quantum mechanics and, 19, 75, 76, 337*n*
quantum nature of light, 57–60, 68, 71
quantum theory and, 320–21
special theory of relativity, 35*n*, 58, 133, 134–78, 181, 183, 187, 219, 290, 318*n*, 344
absolute non-motion and, 151–55
mass-energy relationship, 173–78
Newtonian physics and, 166, 177
principle of constancy of the velocity of light and, 151–55, 164–65
principle of relativity and, 152–55
Pythagorean theory and, 169–73
spacecraft analogy, 156–57
space-time interval, 171–73
Twin Paradox of, 156–57
thought experiments of, 162–64, 181–86, 193–96, 318
time and space, 166–67, 244
Einstein-Podolsky-Rosen (EPR) effect, 252*n*, 316–26, 334*n*
Einstein-Rosen bridges, 207
Einstein's equation, 107*n*, 174–75, 219*n*, 227*n*
Electrodynamics, quantum, 251
Electromagnetic field, 150–51
Electromagnetic force, 230, 251–54, 260–61
Electromagnetic radiation, 53–54, 62, 105, 153, 207
Electromagnetic waves, 62, 318–19

Electron accelerators, 226
Electron cloud, 120–21
Electron diffraction, 107–108
Electron mass, 226–27
Electron volts, 227
Electronic orbits, 57*n*
Electron-positron annihilation, 239, 242
Electrons, 13, 21, 22, 34, 53–54, 226–27
energy levels of, 114, 116
gamma rays and, 124
waves and, 107–109, 110–14
X-rays and, 104
Elementary particle theory, 21
Embrace Tiger, Return to Mountain (Huang), 8
Emissions, 303–306
Emulsion physics, 218*n*
Energy, xxix
frequency and, 63, 104
kinetic, 104, 162, 219, 227, 270
light and, 186
mass and, 35, 136, 173–78, 186, 200, 219–20, 227
mechanical, 176
organic, physics and, 49
quanta and, 106
stellar, 174, 322
thermal, 176
Energy absorption, 57
Energy emission, 57
Energy packets, 56–57
Energy radiation, 53
Enlightenment, 263, 311
physics and, 283–84
Entropy, 246
increasing, 246
maximal, 246
Epilepsy, 42
Epistemology, 288
Equivalence, principle of, 183, 186, 187
Esalen Institute, 4–5, 348
Eta particles, 259
Ether, theory of, 145–46
Euclid, 180*n*, 191, 193, 194

Euclidean geometry, 180n, 191–96, 311
Euclidean space, 171n
European Organization for Nuclear Research (CERN), 313
Event horizon, 206–207
Everett, Hugh, III, 91–92, 94n
Everett-Wheeler-Graham theory. *See* Many Worlds Interpretation of quantum mechanics
Exclusion principle, Pauli, 114, 251–52n
Experience, symbols and, 290, 301, 310
Experimental physics, 285n

Fails, locality, 329
Family numbers, conservation laws of, 270
Feeder accelerator, 226
Feynman, Richard, xxi, 239–43, 262
Feynman diagrams, 239–43, 250, 252, 256–59, 263, 266–67, 275
Feynman perturbation theory, 248n
Fields, matter and, 223
5th Solvay Congress, 40n
Finkelstein, David, xxi, 4, 205, 206, 219, 272, 290, 291–92, 293, 299, 300, 302, 310–11, 348
Finnegans Wake (Joyce), 272
Fission, 175
FitzGerald, George Francis, xix, 148–49, 154–55
FitzGerald-Lorentz contraction, 149
Flower Garland Sutra, 264–66
Forbidden transitions, 302–303
Force
  strong, 252–54, 259–61
  weak, 260–61
  See also Gravity
Ford, Kenneth, 225, 263
Four-dimensional space-time continuum, 168–69, 193, 199
Fourth dimension, 168–69

Frames of reference. *See* Co-ordinate systems
Free will, 28, 336
Freedman, Stuart, xxi, 326, 329, 332, 335
Frequency, 26, 61–63
  energy and, 63, 104
Freud, Sigmund, 43
Friction, 23, 176
Fundamental Physics Group, 335, 336
Fusion, 174–75
  laser, 285

Galaxies, 199, 260
Galaxy clusters, 199
Galilean relativity principle, 138–39, 152, 153
Galilei, Galileo, 6, 12, 24, 26, 31, 119, 138, 140–41, 152, 153, 344
Gamma rays, 124
  electrons and, 124
Gas volume, pressure and, 36–37
Gauge theories, unified, 285n
Gell-Mann, Murray, 272
General theory of relativity, 25, 94, 134, 179–209, 290, 311, 344
Geometry, 190–91, 200
  analytic, 337
  Euclidean, 180n, 191–96, 311
  laws of, 180
Germer, Lester, xx, 107
Ghost waves, 71
Goethe, Johann Wolfgang von, 214
Graham, Neill, 91
Gravitational fields, 185–86, 200
Gravitational mass, 186, 187
Gravitational redshift, 204–205
Graviton, 260–61, 269
Gravity, 182–209, 219n, 260
  acceleration and, 185–86
  light and, 203–205
  Newton's law of, 24–26, 138, 182, 186, 201, 203–204
  subatomic particles and, 260

Great Machine idea, 24, 26, 28, 31, 32, 145, 333
Greatest lower bound, 308
Green light, 56, 59, 62
Guillemin, Victor, 55–56
Gurus, 247

Hafele, J. C., 157n
Hagedorn theory, 246n
Hamilton, W. R., 121
Happiness, 284
Hawking, S. W., 205n
Heart Sutra, The, 268
Heisenberg, Werner, xx, xxxv, 22, 29, 72, 86, 108–109, 121–24, 126, 133–34, 214, 220, 224, 225, 287, 291, 317, 338
Heisenberg's uncertainty principle, 29, 37, 123–26, 233, 248, 256, 261–62, 269, 274, 338
Helium, 174–75
Herbert, Nick, 267n
High-energy particle physics. *See* Particle physics
High-energy photons, 59–60
High-energy virtual photons, 248n
High-frequency light, 62–63
Hinduism, 91, 241, 342, 345–46
Horizon, event, 206–207
Hua Yen Sutra, 266n
Huang, Al Chung-liang, 5, 6, 7–8, 9–10, 348
Hughes, David, 109
Human reactions, chemical reactions and, 50–51
Hydrogen, 12–16, 38–39, 114–17, 120, 174–75, 206, 207
Hydrogen (fusion) bombs, 175
Hydrogen spectrum, 13–16, 39, 57
Hypotenuse, the, 169–71

Idealism, 88
Identity process, 304
Imaginary rest mass, 319n
Increasing entropy, 246
Index of refraction, 142n, 152n

Inertial co-ordinate systems, 138–41, 182–84, 186, 194, 198
Inertial frame of reference. *See* Inertial coordinate systems
Inertial mass, 186–87
Inorganic matter
  organic matter and, 51–52
Inquisition, the, 24
Interactions, 105–106, 251, 252, 268–69
Interference, 68–69, 113, 134
Interferometers, 146, 154–55
Invariance, 167, 200n, 271
Iron, 50
Isotopic spin, 234n

James, William, 41n
Josephson, Brian, 267n, 278n
Joyce, James, 272
Jung, Carl, 33, 69n, 328n

K mesons, 218–19, 240
Kaon decays, 330
Kaots, 259, 273, 278
Keating, R. E., 159n
Kelvin, Lord, 344
Kinetic energy, 104, 162, 219, 227, 270
Knowledge, wisdom and, 337
Koans, 229, 232
Kramers, H. A., xx, 72, 118
Krishna, Lord, 101, 344

La Place, Pierre-Simon, 205n
Lambda, particles, 218–19, 240, 270, 273, 278
Laser fusion, 285
Lattices, 305–10
Lawrence Berkeley Laboratory (LBL), xxxi, 42, 228n, 236, 326, 335, 343
Lenard, Philippe, 58–59
Length
  proper, 155–56
  relative, 155–56
Lepton member, 234n

Leptons, 227–28, 233–34, 235, 271
Light, 33
   double-slit experiment, 66–87,
      92, 105, 110, 314
   energy and, 186
   gravity and, 203–205
   green, 56, 59, 62
   high-frequency, 62–63
   interactions, 105–106
   measuring speed of, 154–55
   the Michelson-Morley
      experiment, 144–50,
      151–52, 153, 154, 164, 344
   particle phenomenon, 105–106,
      133–34
   polarization of, 293–98, 302–10,
      323–26
      quantum mechanics and, 298
      quantum theory and, 324
   quantum nature of, 58–60
   red, 56–57, 59, 62
   velocity of, 62, 135, 136, 142–48,
      151–52, 153–55, 318–19
      principle of the constancy of,
      151–55, 164–65
   violet, 54, 56–57, 59
   visible, 61
   wave phenomenon, 60, 105–106,
      133–34, 318–19
   wave theory of, 65–68
   wave-like, 105
Light quanta, theory of, 133–34
Lobachevsky, Nikolai Ivanovich,
   $180n$
Local causes, principle of, 320, 325,
   329, 331–36
Locality fails, 329
Logic
   classical, 291–93, 298–99, 302,
      306–307, 311
      subatomic particles and, 291–93
   quantum, 285, 291–93, 298, 301,
      302, 306–307, 310
Logical calculus, 288
Logos, mythos and, 290–92
Longchenpa, 312

Lorentz, Hendrik Antoon, xix,
   148–49, 154–55, 165
Lorentz transformations, 148–49,
   155, 160, 164
Low-energy photons, 59–60
LSD, 225, 247
Luther, Martin, $24n$

Macroscopic level, 34, 57
Macroscopic objects, $79n$
Magnetic monopoles, $230n$
Mahayana Buddhism, 266, 267–68
Many Worlds Interpretation of
   quantum mechanics, 92–96,
   333–34, 336
Maps
   space, 236
   space-time, 236–39
Mass, 200, 226–29
   electron, 226–27
   energy and, 35, 136, 173–78,
      186, 200, 219–20, 227
   gravitational, 186–187
   imaginary rest, $319n$
   inertial, 186–87
   relativistic, 226
   rest, 162, 226–27
   velocity and, 161–62
   zero rest, 228
Mass-energy, conservation laws of,
   175, 176, 248, 249, 255,
   261–62, 269, 270
Mass/energy dualism, $219n$
Massless particles, 228–29
Materialism, 88
*Mathematical Foundations of
   Quantum Mechanics, The*
   (von Neumann), 286, 287
*Mathematical Theory of Relativity,
   The* (Eddington), $192n$
Mathematics
   physics and, 4, 17
   subatomic particles and, 289–90
Matrices, 122, 273–78
Matrix mechanics, 122
Matter
   fields and, 223
   inorganic, 51–52
   organic, 51–52

Matter waves, 107, 110, 115, 118, 122
Maximal entropy, 246
Maxwell, James Clerk, 136n, 150
Maxwell's field equations, 150
Maxwell's theory, 133
Measurement
  problem of, 87, 91, 93
  theory of, 87, 338
Mechanical energy, 176
Mechanics
  classical, 150–51, 152
  defined, 20
  laws of, 138–41, 153, 182, 196
  matrix, 122
Meditation, 173, 343
Membrane, one-way, 205–206
Mercury, 201–202
  perihelion of, 201–202, 205
Mesons, 228, 234, 235, 254, 261, 273
  K, 218–19, 240
  pi, 218, 235, 240
Metaphysics, physics and, 91
Michelson, Albert, xix, 144–47, 152
Michelson-Morley experiment, 144–49, 151–52, 153, 154, 164, 344
Microscopic level, 34
Middle Ages, 26, 30
Milky Way galaxy, 208
Miller, Henry, 133
Minkowski, Hermann, xix, 173
Minkowski's flat space-time, 171n
Molecules, 36–37
Momentum, 28, 30, 224, 270
  angular, 231–34
  zero, 270
Momentum-energy space descriptions, 239n
Monopoles, magnetic, 230n
Moon
  the earth and, 27
  motion and, 137
Moon probes, 27
Morley, Edward, xix, 144–47, 152

Motion, 26, 136–42
  absolute, 138, 142, 183
  earth and, 137–39, 145–48
  moon and, 137
  Newton's laws of, 22–23, 25, 26–28, 70
    subatomic particles, 125
  non-absolute, 138, 141–47, 151–53, 184, 209
  sun and, 137, 138
  time dilation and, 155
Muons, 158, 254n
  anti-, 330
Murchie, Guy, 172n
Murphy, Michael, 4
Music of the Spheres (Murchie), 172n
Mystic knowing, 267n
Mythos, logs and, 290–92

Natural philosophy, 30
Nature, discontinuous structure of, 52–53
Negative acceleration, 186–87
Negentropy, 246
Neon, 323
"Neue Bahnen der physikalischen Erkennmis" (Planck), 55n
Neutrinos, 330
Neutron Optics (Hughes), 109
Neutrons, 13, 21, 175, 225, 256–59, 263–64, 270, 273, 277
Newton, Sir Isaac, xxv, 6, 11, 12, 20–21, 22–30, 55, 73, 145, 161, 199, 217, 279, 331, 337, 344
  absolute time, and, 166
  the church and, 23–24, 30
  time and space, 166–68
Newtonian physics, 20–31, 52–55, 73, 110, 122, 161, 166, 168, 177, 205n, 214, 224
  general theory of relativity and, 203–204
  quantum mechanics and, 20–22, 28–31, 34, 37, 44–45, 110, 177, 337–38, 344

Newtonian physics, *(continued)*
    space program and, 27
    special theory of relativity and,
        166, 177
    statistics and, 36–37
Newton's law of gravity, 24–26,
        138, 182, 186, 201, 203–204
Newton's laws of motion, 22–23,
        25, 26–28, 70
    subatomic particles and, 125
Nobel Prize, 14, 33, 58, 107, 345
Nodes, 111–13
Non-being, Being and, 341
Non-inertial co-ordinate systems,
        196
Nonlocal universe, 335
Non-motion, absolute, 141–47,
        151–53, 184, 209
Nuclear particles, 34–35
Nucleons, 253–54, 256, 259
Null process, 304, 305, 306
Numbers
    baryon, 234*n*
    family, conservation laws of, 270
    lepton, 234*n*
    quantum, 114, 217*n*, 234

Objectivity, quantum mechanics
        and, 32–33
Observables, 77–78
One-way membrane, 205–206
Ontology, 288
Oppenheimer, 205*n*
Optics, 53
Optiks (Newton), 11
Organic energy, physics and, 49
Organic matter, inorganic matter
        and, 51–52
Oscillators
    atomic, 53–54, 55
    quantized, 55
Oxygen, 50

Paradigms, 285
Parallel axiom, 180*n*
Particle accelerators, 215, 217–18,
        219, 226

Particle phenomenon, light and,
        105–106, 133–34
Particle physics, 18, 34–35, 122,
        158, 174, 176–77, 188*n*,
        213–78, 347–48
    basic questions of, 220, 231
    Buddhism and, 262–63, 264–66
    dynamics of, 313
    hardware of, 217
    purpose of, 217
    quantum field theory and, 220
    relativity and, 219
    subatomic particles, 219–20,
        224–25
    tools of, 217
Particle theory, elementary, 21
Particles
    D, 228*n*
    defined, 51–52
    eta, 259
    lambda, 218–19, 240, 270, 273,
        278
    massless, 228–29
    nuclear, 34–35
    standing waves and, 118
    Tau, 228*n*
    W, 261
    See also Subatomic particles; types
        of particles
Pauli, Wolfgang, xx, 33, 114
Pauli exclusion principle, 114,
        251–52*n*
Penrose, Roger, 205
Perturbation theory, 248*n*
Phenomena
    psychic, 331
    quantum, 313–14
*Philosophiae Naturalis Principia
    Mathematica* (Newton),
        24–25, 27
Philosophy, natural, 30
Photoelectric effect, 58–60, 63, 105
Photography, time-lapse, 50
Photomultiplier tubes, 324–25
Photon detectors, 82–83. *See also*
        Photomultiplier tubes

Photons, 22*n*, 39, 58–60, 63,
68–71, 74–87, 92, 143–44,
206*n*, 218*n*, 227, 233–35,
238, 239–40, 242, 243*n*,
243–44, 247–51, 261–63,
271, 286–87, 298, 314,
323–24
high-energy, 59–60
high-energy virtual, 248*n*
low-energy, 59–60
as organic, 70
virtual, 247–55, 261–62
Physical concepts, defined, 9
Physics, 3–4, 96, 131–32
bubble chamber, 218*n*
Buddhism, and, 311
the church and, 23–24
classical, xxv, 53–54, 150–51,
289–90
quantum mechanics and,
94–96, 124–26
consciousness and, xxviii, xxx, 88,
92–93, 96
counter, 218*n*
defined, 201
emulsion, 218*n*
enlightenment and, 283–84
experimental, 285*n*
mathematics and, 4, 17
metaphysics and, 91
organic energy and, 49
pre-relativity, 144*n*
psychology and, 33–34, 180
quantum mechanics and, 19, 221
theoretical, 272, 285*n*
See also Newtonian physics:
Particle physics
Physics/Consciousness Research
Group, 4
Pi mesons, 218, 235, 240
Pions, 158, 254–59, 263–64, 270,
273, 275–77, 330
virtual, 255–57, 259
Planck, Max, xix, 52–59, 63, 72,
104, 106–107, 118, 124,
142, 214, 230, 347
Einstein and, 59, 72

quantum of action, 233, 344
Planck's constant, 56, 58*n*, 63, 233
Bohr and, 233
color and, 56
Einstein and, 233
quantum mechanics and, 63
subatomic particles and, 233
Planck's equation, 107*n*
Plato, 159, 160
Podolsky, Boris, xx, 316–21
Polarizers, 293–98, 302–306,
323–25
diagonal, 296–98, 302, 304–305
horizontal, 294–98, 302–303,
305
vertical, 294–98, 302–303, 305
Positive acceleration, 186–87
Positrons, 234–35, 243*n*, 243–44,
251, 271
Potentia, 72, 287
Pragmatism, 41, 79*n*
quantum mechanics and, 41, 79*n*,
88–89, 90, 109–10, 251–52
Prajnaparamita Surras, 267–68
Pre-relativity physics, 144*n*
Pressure
gas volume and, 36–37
sea level, 36
Price, Dick, 4
Probability, quantum mechanics and,
29, 74–75, 81–82, 86, 88,
91, 110, 117–18
Probability waves, 72, 117
Problem of measurement, 87, 91,
93, 338
Progress, science, and, 101–102
Proof, scientific, 302
Proper length, 155–56
Proper time, 155–57, 162
Proportionality, constant of, 63
Protons, 12–13, 21, 218, 225, 226,
234, 236, 240, 252–59,
263–64, 270, 273–76
anti-, 263, 264
Psychic phenomena, 331
Psychology, 288
physics and, 33, 180

Ptolemaic astronomy, 217n
Pythagoras the Greek, 169
Pythagorean theorem, 169–72

Quadruplets, 306
Quanta, energy and, 106
Quantification, 31–32
Quantized oscillators, 55
Quantum, defined, 20
Quantum electrodynamics, 251
Quantum field theory, 145n,
    221–24, 239, 242–43, 244,
    248, 251, 252
  particle physics and, 220
  quantum mechanics and, 222
Quantum jump, 83, 116n
Quantum logic, 285, 291–93, 298,
    301, 302, 306–307, 310
Quantum mechanics, xxv–xxvi, 18,
    79–80, 214, 251–52n, 313,
    317, 330
  basic paradox, 225
  Bohr on, 28, 125–26, 289, 337n
  classical physics and, 94–96,
    125–26
  coherent superposition and,
    299–300
  common sense contradictions and,
    29
  Compton and, 103–105
  the Copenhagen Interpretation of,
    40–44, 76, 88, 92, 93,
    94–95, 317, 336, 347
  defined, 20
  Eastern literature and, 101
  Einstein and, 19, 75, 76, 337n
  major contribution of, 223–24
  Many Worlds Interpretation of,
    92–96, 333–34, 336
  Newtonian physics and, 20–22,
    28–31, 34, 37, 44–45, 110,
    177, 337–38, 344
  objectivity and, 32–33
  the observed system, 76–78, 84
  the observing system, 76–77, 84
  philosophical implication of, 52

physics and, 19, 221
Planck's constant and, 63
the polarization of light and, 298
pragmatism and, 41, 79n, 88–89,
    90, 109–10, 251–52
preparation and measurement in,
    75–85
probability and, 29, 74–75,
    81–82, 86–88, 91, 110,
    117–18
the problem of measurement in,
    87, 91, 93
quantum field theory and, 222
reality and, 33
relativity and, 222
Schrödinger wave equation and,
    117
Schrödinger's cat and, 94–96
the split-brain analysis, 42–44
statistics and, 37–40
as the study of correlations
    between experiences, 110
subatomic particles and, 34–40,
    51, 96–97, 117, 220–22, 322
von Neumann and, 286–88
the wave function, 88–94, 95–96,
    119
wave-particle duality and, 70–73,
    106–107, 133–34
Quantum numbers, 114, 217n, 234
Quantum paradox, basic, 103n
Quantum phenomena, 313–14
Quantum theory, xxv, xxviii, xxx, 20,
    217, 264, 288, 289–90, 299,
    310–11, 317–18, 347
  Bell's theorem and, 322–23,
    325–26, 329, 331, 334–35
  Einstein and, 320–21
  Einstein-Podolsky-Rosen effect
    and, 316–26
  Planck's quantum of action and,
    233, 344
  polarization of light and, 324
Quantum topology, 311, 348
Quarks, 230n, 272, 285
Quasars, 208

Rabi, Isidor I., 10, 345
Radiation
  black-body, 54–55, 106, 233, 344
  cosmic, 158
  electromagnetic, 53–54, 62, 105, 153, 207
  energy, 53
Radii, 193–96
Radio waves, 62, 318–19
Radium, 38
Rate of spin, 231–32
Rauscher, Dr. Elizabeth, 335
Reactions, chain, 175
Reactors, atomic, 175
Reality, xxviii, 94, 96, 97, 343, 344
  quantum mechanics and, 33
Red light, 56–57, 59, 62
Redshift, gravitational, 204–205
Reference, frames of. *See* Co-ordinate systems
Refraction, index of, 142*n*, 152*n*
Relative length, 155–56
Relative time, 155–59
Relativistic mass, 226
Relativity, xxiii, 18
  Eastern literature and, 101
  Galilean principle, 138–39, 152, 153
  general theory of, 25, 94, 134, 179–209, 290, 311, 344
    astronomy and, 201–204
    black hole theory, 205–209
    equations of, 200
    gravity and acceleration, 186–88
    Newtonian physics and, 203–204
    principle of equivalence, 183, 186, 187
  particle physics and, 219
  quantum mechanics and, 222
  special theory of, 35*n*, 58, 133, 134–78, 181, 183, 187, 219, 290, 318*n*, 344
    absolute non-motion and, 151–53

mass-energy relationship, 173–78
Newtonian physics and, 166, 177
principle of constancy of the velocity of light and, 151–55, 164–65
principle of relativity and, 152–55
Pythagorean theory and, 169–72
spacecraft analogy, 156–57
space-time interval, 171–73
Twin Paradox of, 156–57
Religions
  Eastern, 342–48
  science and, 97
Rest mass, 162, 226–27
Riemann, Georg Friedrich, 180*n*
Right triangles, 169–71
Rosen, Nathan, xx, 316–21
Roshi, Baker, 132
Roshi, Suzuki, 132
Russell, Bertrand, 109, 189
Rust, 50
Rutherford, Ernest, 13

Sarfatti Jack, 267*n*, 328*n*
Scattering Matrix theory. *See* S-Matrix theory
Schilpp, Paul, 262*n*
Scholasticism, 26
Schrödinger, Erwin, xx, xxxv, 110–19, 121–22, 286, 318
Schrödinger wave equation, 77, 80–94, 113–18, 122, 287, 300
  quantum mechanics and, 117
Schrödinger's cat, 94–96
Science
  progress and, 101–102
  religion and, 97
  the self and, 101–103, 127
  subatomic events and, 127
Scientific "proof," 302
Scientific "truth," 41, 42, 302

Scientists, technicians and, 10–11, 16–17
Sea-level pressure, 36
Seat of the Soul, The (Zukav), xxx
Self, the, science and, 101–103, 127
Self-interaction, 263
Shiva, 241
Singularity, black hole, 206–207
Slater, John, xx, 72, 117
S-Matrix theory, 115n, 122, 222n, 272–78
    subatomic particles and, 273–77
Smith, Dr. Felix, 232n
Snyder, S., 205n
Sodium, 11–12, 50, 204–205
Sodium chloride, 50
Sodium spectrum, 11–12
Solar systems, 199, 208–209, 260
    movement of, 189
Solitons, 285n
Sommerfield, Arnold, 14
Sound, velocity of, 165
Sound waves, 318
Space
    Euclidean, 171n
    time and, 166–78, 196, 310–11
Space maps, 236
Space program, Newtonian physics and, 27
Space-like separations, 164n, 319
Space-time continuum, 188–90, 197–201, 209
    four-dimensional, 168–69, 193, 199
Space-time diagrams, 242, 243–44, 274
Space-time interval, 171–73
Space-time maps, 236–39
Space-time pictures, 244–45
Special theory of relativity, 35n, 58, 133, 134–78, 181, 183, 187, 219, 290, 318n, 344
Spectrum, 38–39
    hydrogen, 13–16, 39, 57
    sodium, 11–12
    white-light, 11–12, 15–16
Spheres, 191

Spherical trigonometry, 199
Spin, 234n, 235, 270
    isotopic, 234n
    rate of, 230–32
    zero, 314–16
Standing waves, 110–17
    particles and, 118
Stanford Linear Accelerator Center (SLAC), 228n
Stapp, Henry Pierce, xxi, 42, 69, 78–80, 88, 89, 90, 322, 326, 328, 331, 332, 335, 336, 345
Star systems, 199
    binary, 207
Starlight deflection, 203–204
Statistics
    Newtonian physics and, 36–37
    quantum mechanics and, 37–40
Stellar energy, 174, 322
Stern-Gerlach device, 315–17, 321
Strangeness, 234n
Strong force, 252–54, 259–61
Subatomic particles, 175, 213–79, 289
    angular momentum of, 231–34
    Bohm's hypothesis and, 339–43
    characteristics of, 225–34
    classical logic and, 291–93
    conservation laws, 268–71
    gravity and, 260
    mathematics and, 289–90
    Newton's laws of motion and, 125
    particle physics and, 219–20, 224–25
    Planck's constant and, 233
    quantum mechanics and, 34–40, 51, 96–97, 117, 220–22, 322
    S-Matrix theory and, 273–77
    time flow and, 245–47
Subatomic phenomena, 21–22, 29, 35, 38, 40, 41n, 57
    Born and, 117
Sun, motion and, 137, 138
Superdeterminism, 333
Supergravity theories, 260n
Superluminal information transfer theory, 326–28, 333

Superluminal quantum connectedness, 331
Superluminal speed, 319–20
Superposition, coherent, 299–301
Sutras, 264–66, 267–68
Symbols, 343
    experience and, 290, 301, 310
Symmetries, 271
Symmetry, laws of, 177
Synchronicity, 69n, 328n

Tachyons, 319n
Tai Chi Master, 5, 7–8, 348
Tantra, 346–47
Tantric Buddhism, 220–21, 346–47
Taoism, 201, 266n
Tao particle, 228n
Technicians, scientists and, 10–11, 16–17
Telepathy, 331
Terrell, James, 159–61
Theoretical physics, 271–72, 285n
Theory of measurement, 87, 91, 93, 338
Thermal energy, 176
Thermodynamics, second law of, 246
Thomas, Saint, 97
Thought experiments, 162–64, 181–86, 193–96, 318
Tibetan Buddhism, 312, 343, 347
Time
    absolute, 166
    proper, 155–57, 162
    relative, 155–59
    space and, 166–78, 196, 310–11
    universal, 162
Time dilation, 158–161, 165
    motion and, 155
Time irreversability, 246n
Time-lapse photography, 50
Time-like separation, 164n
Time reversability, 246n
Time-reversal invariance principle, 271
Topology, quantum, 311, 348
Torsion, 200n

Transformation laws, classical, 140, 152, 153–54, 164–65
Transition tables, 303–305, 307
Transitions
    allowed, 302
    forbidden, 302–303
Triangles, 191–92
    right, 169–71
Trigonometry, spherical, 199
Triplets, 306
Truth
    absolute, 41–42
    scientific, 41, 42, 302
Tubes, photomultiplier, 324–25
Twin Paradox of the special theory of relativity, 156–57
Two-particle systems, 314–16

Uhura, 207
Ultra-violet catastrophe, 54
Uncertainty principle, Heisenberg's, 29, 37, 123–26, 233, 248, 256, 261–62, 269, 274, 338
Unconscious, the, 43
Unified gauge theories, 285n
Universal time, 162
Universe, nonlocal, 335
Uphill interactions, 270
Uranium, 120

Vacuum diagrams, 267–68
Velocity, 26, 61–62, 158, 224
    of light, 62, 135, 136, 142–48, 151–52, 153–55, 318–19
    principle of the constancy of, 151–55, 164–65
    mass and, 161–62
    of sound, 165
Violet light, 54, 56–57, 59
Virtual particle theory, 251–52
Virtual photons, 247–55, 261–62
Virtual pions, 255–57, 259
Vishnu, 241
Visible light, 61
Volts, electron, 227
Volume
    gas, pressure and, 36–37
    zero, 206

Von Neumann, John, xx, 232*n*, 286,
    292–93, 302, 310, 311, 338
  quantum mechanics and, 286–88

W particle, 261
Walker, E. H., 69
Watts, Alan, 8
Wave equation, Schrödinger, 77,
    80–94, 113–18, 122, 287,
    300
Wave function, 80–85, 88, 89,
    91–92, 285–88, 300
  collapse of, 82–85, 95–96,
    330–31*n*
  quantum mechanics and, 88–94,
    95–96, 118
Wave mechanics, 63, 67–68
Wave-particle duality, 70–72,
    103–107, 133–34, 223–24
  quantum mechanics and, 70–73,
    106–107, 133–34
Wave theory of light, 65–68
Wavelengths, 61–63
Waves, 60
  amplitude, 61
  electromagnetic, 62, 318–19
  electrons and, 107–109, 110–14
  frequency, 61
  ghost, 71
  matter, 107, 110, 115, 118, 122
  probability, 72, 117
  radio, 62, 318–19
  sound, 318

  standing, 110–17
    particles and, 118
    velocity, 61–62
Weak froce, 260–61
Weizenbaum, Joseph, 26
Wheeler, John, 31, 91–92
White holes, 208
White-light spectrum, 11–12, 15–16
Wisdom, knowledge and, 337
Wonder, subjective experience of, 44
World of Elementary Particles, The
    (Ford), 225*n*, 263
Wormholes, 207
Wu Li, defined, 5–7

X-rays, 61
  electrons and, 104

Yin and yang, 43, 177
Yogis, 247
Young, Thomas, xix, 60, 66–68,
    106, 110
Young's double-slit experiment,
    66–71
Young's wave theory, 71
Yukawa, Hideki, xx, 252, 253–54,
    257

Zen Buddhism, 132, 229
Zen Mind, Beginner's Mind (Roshi),
    132
Zero momentum, 270
Zero rest mass, 228
Zero spin, 314–16
Zero volume, 206

THE GREEK ALPHABET

A	α	ALPHA	N	ν	NU
B	β	BETA	Ξ	ξ	XI
Γ	γ	GAMMA	O	o	OMICRON
Δ	δ	DELTA	Π	π	PI
E	ϵ	EPSILON	P	ρ	RHO
Z	ζ	ZETA	Σ	σ	SIGMA
H	η	ETA	T	τ	TAU
Θ	θ	THETA	Υ	υ	UPSILON
I	ι	IOTA	Φ	φ	PHI
K	κ	KAPPA	X	χ	CHI
Λ	λ	LAMBDA	Ψ	ψ	PSI
M	μ	MU	Ω	ω	OMEGA

FOOTNOTES FOR STABLE PARTICLE TABLE

1.   This table was compiled with the assistance of the Particle Data Group, Lawrence Berkeley Laboratory, Berkeley, California. According to their convention, stable particles are particles that do not decay by strong interaction; but they do decay by electromagnetic and weak interactions. In fact, (as the table shows), the majority of stable particles are not 'stable' in the usual sense.

2.   Incredible as it may seem, physicists measure particle lifetimes (and masses) to far greater degrees of accuracy than indicated here. ("Review of Particle Properties," Physics Letter, 75B, 1, 1978.) (updated bi-annually).

3.   Particles with this footnote are speculative to one degree or another. The graviton has been predicted solely on a theoretical basis while the tau neutrino, the F particle, and the charmed lambda have some experimental evidence to support their existence. However, none of them have been accepted, at this date, as confirmed particles by the Particle Data Group, Lawrence Berkeley Laboratory, the internationally accepted compiler of particle data. (The parameters shown for these particles have not been measured, but they generally are assumed to have the values shown). The blank spaces in the table represent a lack of data. [Information on the Tau and D particles, the newest particles as this table goes to print (1978), is still incomplete.]

4.   The neutral kaon has two average lifetimes. If it decays in the shorter time it is called a K_S^0 ("K zero short"). If it decays in the longer time it is called a K_L^0 ("K zero long"). All the other particles have only one average lifetime.

STABLE PARTICLE

	PARTICLE NAME	SYMBOL	MASS	SPIN	CHARGE
	PHOTON	γ	0	1	NEUTRAL
	GRAVITON [3]	—	0	2	NEUTRAL
LEPTONS	ELECTRON NEUTRINO	ν_e	0	½	NEUTRAL
	ELECTRON	e^-	1	½	NEGATIVE
	MUON NEUTRINO	ν_μ	0	½	NEUTRAL
	MUON	μ^-	207	½	NEGATIVE
	TAU NEUTRINO [3]	ν_τ	—	½	NEUTRAL
	TAU	τ^-	3536	½	NEGATIVE
MESONS	PION	π^+	273	0	POSITIVE
		π^-	273	0	NEGATIVE
		π°	264	0	NEUTRAL
	KAON	K^+	996	0	POSITIVE
		K°	974	0	NEUTRAL
	ETA	η	1074	0	NEUTRAL
	D	D^+	3656	0	POSITIVE
		D°	3646	0	NEUTRAL
	F [3]	F^+	—	0	POSITIVE
BARYONS	PROTON	p	1836	½	POSITIVE
	NEUTRON	n	1837	½	NEUTRAL
	LAMBDA	Λ	2183	½	NEUTRAL
	SIGMA	Σ^+	2328	½	POSITIVE
		Σ°	2334	½	NEUTRAL
		Σ^-	2343	½	NEGATIVE
	XI	Ξ°	2573	½	NEUTRAL
		Ξ^-	2586	½	NEGATIVE
	OMEGA	Ω^-	3272	3⁄2	NEGATIVE
	LAMBDA [3]	Λ_c^+	—	—	POSITIVE

TABLE[1]

ANTI PARTICLE	TYPICAL MODE OF DECAY	AVERAGE LIFETIME [2]
SAME PARTICLE	STABLE	INFINITE
SAME PARTICLE	STABLE	INFINITE
$\bar{\nu}_e$	STABLE	INFINITE
e^+ (positron)	STABLE	INFINITE
$\bar{\nu}_\mu$	STABLE	INFINITE
μ^+	$\mu^- \longrightarrow e^- + \bar{\nu}_e + \nu_\mu$	2.2 MILLIONTHS OF A SECOND (2.2×10^{-6})
$\bar{\nu}_\tau$	——	——
τ^+	$\tau^- \longrightarrow e^- + \bar{\nu}_e + \nu_\tau$	——
π^- } SAME AS THE	$\pi^+ \longrightarrow \mu^+ + \nu_\mu$	26 BILLIONTHS OF A SECOND (26×10^{-9})
π^+ } PARTICLES	$\pi^- \longrightarrow \mu^- + \bar{\nu}_\mu$	26 BILLIONTHS OF A SECOND (26×10^{-9})
SAME PARTICLE	$\pi^\circ \longrightarrow \gamma + \gamma$	80 QUINTILLIONTHS OF A SECOND (80×10^{-18})
K^-	$K^+ \longrightarrow \mu^+ + \nu_\mu$	12 BILLIONTHS OF A SECOND (12×10^{-9})
$\bar{K}^\circ$	$K^\circ_{SHORT} \longrightarrow \pi^+ + \pi^-$ $K^\circ_{LONG} \longrightarrow \pi^+ + \pi^- + \pi^\circ$ [4]	90 TRILLIONTHS OF A SECOND (90×10^{-12}) 52 BILLIONTHS OF A SECOND (52×10^{-9})
SAME PARTICLE	$\eta \longrightarrow \gamma + \gamma$	0.8 QUINTILLIONTH OF A SECOND (0.8×10^{-18})
D^-	$D^+ \longrightarrow K^- + \pi^+ + \pi^+$	——
$\bar{D}^\circ$	$D^\circ \longrightarrow K^- + \pi^+$	——
F^-	——	——
$\bar{p}$	STABLE	INFINITE
$\bar{n}$	$n \longrightarrow p + e^- + \bar{\nu}_e$	918 SECONDS
$\bar{\Lambda}$	$\Lambda \longrightarrow p + \pi^-$	0.3 BILLIONTHS OF A SECOND (0.3×10^{-9})
$\bar{\Sigma}^-$	$\Sigma^+ \longrightarrow p + \pi^\circ$	80 TRILLIONTHS OF A SECOND (80×10^{-12})
$\bar{\Sigma}^\circ$	$\Sigma^\circ \longrightarrow \Lambda + \gamma$	58 SEXTILLIONTHS OF A SECOND (58×10^{-21})
$\bar{\Sigma}^+$	$\Sigma^- \longrightarrow n + \pi^-$	0.2 BILLIONTH OF A SECOND (0.2×10^{-9})
$\bar{\Xi}^\circ$	$\Xi^\circ \longrightarrow \Lambda + \pi^\circ$	0.3 BILLIONTH OF A SECOND (0.3×10^{-9})
$\bar{\Xi}^+$	$\Xi^- \longrightarrow \Lambda + \pi^-$	0.2 BILLIONTH OF A SECOND (0.2×10^{-9})
$\bar{\Omega}^+$	$\Omega^- \longrightarrow \Xi^\circ + \pi^-$	0.1 BILLIONTH OF A SECOND (0.1×10^{-9})
$\bar{\Lambda}_c^-$	——	——

An Invitation

At the heart of *The Dancing Wu Li Masters* (and every inquiry into quantum physics), in my opinion, is the inviting possibility of a relationship between consciousness and physical reality. That relationship, for me, has become more and more evident since I wrote *The Dancing Wu Li Masters*. It has become the foundation of my life and my deepest joy. I believe that this relationship is also becoming evident for millions of other individuals.

In 1989 I wrote *The Seat of the Soul,* about evolution, the soul, and authentic power—the alignment of the personality with the soul. If these topics attract you, I invite you to join me by visiting www.zukav.com or www.seatofthesoul.org or by writing to me at:

> The Seat of the Soul Foundation
> PO Box 339
> Ashland, OR 97520

> With Joy,
> Gary Zukav